INTERSECTING Sets

INTERSECTING Sets

A Poet Looks at Science

ALICE MAJOR

THE UNIVERSITY OF ALBERTA PRESS

Published by

The University of Alberta Press
Ring House 2
Edmonton, Alberta, Canada T6G 2E1
www.uap.ualberta.ca

LIBRARY AND ARCHIVES CANADA CATALOGUING IN PUBLICATION

Major, Alice
    Intersecting sets : a poet looks at science / Alice Major.

Includes bibliographical references and index.
Issued also in electronic format.
ISBN 978-0-88864-595-1

    1. Major, Alice. 2. Literature and science. 3. Poetry. 4. Poets, Canadian (English)—
20th century. I. Title.

PS8576.A515Z465 2011   C811'.54   C2011-906296-8

First edition, first printing, 2011.
Printed and bound in Canada by McCallum Printing Group, Edmonton, Alberta.
Copyediting and Proofreading by Meaghan Craven.
Indexing by Judy Dunlop.

The University of Alberta Press is committed to protecting our natural environment. As
part of our efforts, this book is printed on Enviro Paper: it contains 100% post-consumer
recycled fibres and is acid- and chlorine-free.

The University of Alberta Press gratefully acknowledges the support received for its
publishing program from The Canada Council for the Arts. The University of Alberta Press
also gratefully acknowledges the financial support of the Government of Canada through
the Canada Book Fund (CBF) and the Government of Alberta through the Alberta
Multimedia Development Fund (AMDF) for its publishing activities.

Canadä Canada Council   Conseil des Arts   Government
        for the Arts   du Canada   of Alberta ■

Contents

Acknowledgements

THIS BOOK WOULD HAVE BEEN IMPOSSIBLE without grants
from the Alberta Foundation for the Arts and the Edmonton
Arts Council, which gave me time to complete the manuscript.

It would have been even more impossible without all the
other writers, from Martin Gardner on, who have translated science into words a poet could understand. Writers
like John Gribbin and Brian Greene have made cosmology
and mathematics as magical to me as Antonio Damasio, Ellen
Dissanayake, Stephen Pinsky and Richard Dawkins have made
the human and cognitive side. I owe an immeasurable debt
to all the writers listed in the bibliography, and to others who
aren't specifically referenced in the book but whose work has
informed my understanding.

As always, I am immensely indebted to my dear husband,
David Berger, who not only had the shelves built to hold all
those books, but also puts up with living in the welter of paper
that seems to come with having a writer in the house.

I would also like to thank rose researcher Dr. Neville Arnold and Edmonton composer Don Ross for their specific insights and assistance. Thanks also to the people who helped to locate or generate the images used to illustrate the text: Robert Vaughan Moody (University of Victoria), Dirk Frettlöh (Universität Greifswald) and Clinton Curry (Stony Brook University).

The quotes from Derek Walcott, Shirley Serviss, Rhona McAdam and Mimi Khalvati are used with the kind permission of those poets.

Finally, thank you to the University of Alberta Press and its staff who have gallantly taken on production of this book after publishing three volumes of my poetry. They have always been wonderful to work with.

And thanks, universe.

The Magpie's Eye

Why write a book of essays about poetry and science? Surely the intersection of these two fields must be virtually an empty set, or at least one with very few members. Writing poetry seems like such a marginal activity in these days when people enjoy so many other options for communicating. Writing about poetry and science will surely attract fewer fans than a garage band with rap lyrics accompanied by accordion-players in lederhosen.

However, interesting things happen at edges—at boundaries, membranes, seacoasts and the circumference of a convergence circle in the complex number plane. The rap-accordion band could go platinum in our fusion-loving days. (Probably there's one being formed as I write this.) I don't expect fusion for this book, but it seems to me that poets and scientists have much to say to each other and should take the plugs out of their ears.

A STORY: a conference on arts policy. Rows of chairs are set out in the upper lobby of Edmonton's Winspear Centre, a gracious contemporary concert hall. There's the table at the front behind which four panellists are lined up like judges at an assizes. There are the obligatory microphones and a shuffling trail of witnesses who want to testify. In the usual way of such events, the dialogue resembles particles popping in and out of existence in a cloud chamber—remarks or questions from the audience members are directed at some point made several speakers/questions back. A smooth flow doesn't emerge, though there is a random transfer of energy.

I was there waiting for my turn to get to the mike. A comment by one of the speakers had made me think of an article I had recently read in *The Sciences*—an analysis of Jackson Pollock's abstract "drip" paintings that demonstrated how intensely fractal his work is. His patterns scale similarly all the way from the smallest areas of the canvas to the largest. Pollock's patterns represent a kind of relationship that gives humans much pleasure—it's the way small patches of cloud are similar in pattern to large ones, or small ripples of water to large waves. When I gain the mike, I want to say that discoveries like this one about Jackson Pollock are important to all artists, as well as to arts policy-makers. Understanding the processes that drive the human mind to explore and enjoy is essential to connecting with an audience.

But I explained it badly. The panellists looked at me as if I were one of those eccentrics who pop up at public forums and ask about alien abduction. One of them responded by saying something like, "I don't need to understand how my brain works when I'm creating, any more than I need to understand how my computer works."

I wanted to wrest the mike back from the person who had been standing patiently behind me and shout, "It does matter! And if that's how you think about your brain, even your metaphors are out of date."

But the chatter had zoomed off to another part of the cloud chamber, to bemoan the lack of federal funding for the arts. I took my seat meekly. But the idea for this book was born back then, and I have been moving toward it in a crabwise fashion ever since.

A LITTLE HISTORY about the relationship between scientists and poets: we weren't always so divided. The great age of Western science, which began around the time of Bacon, flowered into Newton and Leibniz, and surged on into the Enlightenment, was a great inspiration to poets. Writers like John Donne could be deeply intrigued by the discoveries of scientists and explorers and by their philosophical implications. Poets like Alexander Pope could write lines like "God said 'Let Newton be' and all was light" and celebrate the great, lawful, intricate and orderly world that mathematics seemed to lay out. Writers of the eighteenth century weren't worried about reductionism, about reducing the universe to a few fundamental equations. They liked a framework that allowed for explanations.

Things began to shift with the mystical poet, William Blake, and then with the Romantics. Science became suspect—the source of "dark, Satanic mills" and the social ills resulting from the Industrial Revolution. Science, for the nineteenth-century poet, was against nature, not for it. (Although a Romantic writer like Coleridge could attend scientific lectures "to replenish my store of metaphors.") Into the twentieth century, science kept getting bigger—a great, lumbering unhuman beast that emitted nuclear warheads and uncertainty principles.

This idea of the two cultures isn't new—I apologize for hashing it over—but it's worth reminding ourselves how thoroughly it came to dominate the intellectual landscape. In 1962 British poet-laureate-to-be Ted Hughes could write: "The Scientific Spirit has bitten so many of us in the nape and pumped us full of its eggs, the ferocious virus of abstraction." Robert Graves could pontificate: "Experimental research is all very well for a scientist. He carries out a series of routine experiments in the properties (say) of some obscure metallic compound...But poetry cannot be called a science; science works on a calm intellectual level, with proper safeguards against imaginative freedom."[1]

THIS CHASM between disciplines shaped those of us who grew up in the 70s and 80s, and can still be a habit of mind even among younger people in academic silos. Of course, the divide was never all that rigid. Poets swooped around its crags, occasionally diving in for scavenged metaphors. Occasionally, august nods were exchanged from cliff to cliff, as when Guadalupe poet Saint John Perse, accepting the Nobel Prize in Literature in 1960, could say that science and poetry should not be considered hostile brothers, "for they are exploring the same abyss and it is only in their modes of investigation that they differ."

By the second half of the twentieth century, an increasing amount of "popular" science writing was being published, as scientists made a valiant effort to translate their worlds into something the general public could understand. In the last decade, research has been pouring in on various aspects of cognitive and evolutionary science, and this too is finding its way into accessible summary form.

I came across one such book at an impressionable age. My parents bought my younger sister a copy of *Relativity for the Millions* by Martin Gardner as a Christmas present, back in the early 1960s. Perhaps it came the same Christmas I was given my chemistry set—a large red tin box that opened like a thick book to display a nested microscope and rows of test tubes filled with substances like diatomaceous earth and a loath-some fly floating in clear liquid. I cracked a few glass slides by grinding the lens too far down into their faces, looked at strands of dog hair and bits of leaf, and never opened the test tube with the fly in it. Then the box ended up under the bed until my little brother took it out and became a scientist.

But that book! I devoured it. It was a mysterious gift. I can't remember why my mother would have thought it suitable for an eleven-year-old. But we were a modest household, and I accepted any book, however random, with interest. The dust jacket had an illustration in blues and purples, in which a spiral galaxy and a distorted clock floated. As I ruffled the pages, the same blue reappeared in frequent pictures: a man in a fedora watching a train pass. A clock beside a tipping hourglass. An odd saddle-shaped object covered in a grid of curving lines like distorted graph paper. It seemed like a tale of strange dimensions, a narrative of some urban Narnia.

When I read the text, it became even stranger than Narnia. I was introduced to geometries that differed wildly from the four-square certainty of right angles, as though I had been ordered to wrap a planet (or that saddle shape) in a flat sheet of Christmas wrapping paper. I was asked to imagine being in an elevator in the darkest emptiest region of space, free-floating. And then to imagine what would happen if an immensely long rope started to tow that elevator cubicle through space. I was invited to imagine travelling in a space-ship as fast as a photon.

I fell into this strange geometry, as enchanted as if I had fallen into one of the pools in The Wood Between the Worlds. This book became the basis of a lifelong fascination with physics and cosmology. Odd, perhaps, for a child who was not particularly good at arithmetic and who grew up to be a poet, but Gardner's elegant translation of mathematics into text turned out to be the most wonderful and formative of Christmas gifts.

So years later, in university English classes, I never understood the faintly sneering description of all science as reductive. All a scientist is trying to do is tell a story of how things came to be, just as any writer is. How could telling such amazing narratives about the universe be a reduction? What on Earth could we lose by understanding them?

Scientists and poets have often been stamped as antithetical, but they share this: science and poetry are both considered to be somewhat marginal forms of human activity, elite, not ordinarily needed by most people for their day-to-day functions. Nevertheless, both are central to understanding how human beings fit into the world. What scientists and poets do truly is "for the million."

HAVING DECIDED TO WRITE THIS BOOK, I soon realized that my big problem was how to tackle it. How was I going to organize subjects that ranged from relativity theory to evolution? And I soon found I was facing the same question that's central to fiction or poetry: who is speaking, and to whom?

First of all, I'm not a scientist; nor am I a science writer. I have read many popular science books, but I can't claim a technical background. I quickly realized there was no point trying to duplicate what many people with better qualifications are doing. I cannot teach people about cognitive science or chemistry. Nor am I a literary critic or philosopher in aesthetics. Once again, I don't have the credentials—I can never remember what Saussure or Lévi-Strauss said without looking it up. I have spent the last few months realizing how I could easily get things *very* wrong.

I'm just a working poet, trying to understand what I'm doing and why, partly because that's the sort of question that comes at you when you get older. But also because I really wasn't happy with the theories I had been submerged in when I went to university thirty years ago. Perhaps it's just that I have always felt a little out of step with literary fashion and want to justify what I've been doing all these years. However, I think that literary theories of the twentieth century did have a frequently unfortunate influence on a generation of poets—in particular the ideas that parade down the intellectual boulevard under the parasol of postmodernism.

Postmodernism, of course, became a roomy and many-ribbed cover for varied thinkers to cluster under. Some of its concepts delivered a necessary poke from the brolly, such as the idea that "masterpieces" should be reread in the light of their hidden assumptions about power and gender. However, I actively disliked many of its sub-isms and ultimatums. The more I read about science, the more I felt that much of post-modernism was based on fossilized ideas from and about science, while science itself *had* changed profoundly and creatively, and had found useful things to say. Like science writer Steven Johnson, I think that if Marshall McLuhan is right and media are extensions of our central nervous system, then we need a theory of the central nervous system as much as we need a theory of media—in other words, our theories about how arts and media work need to be tied back to real human beings and how *they* work.[2]

So in this book, I ended up as a magpie—such a conspicuous bird here on the northern plains—picking up bright, oddly shaped ideas that attracted me from various disciplines and arranging them along with anecdotes that I hope give them context. I hope I can offer some insights on the dialogue between art and science from a poet's sparsely populated stretch of the fusion zone. I hope you find it interesting. There are no lederhosen.

That Frost Feeling

ONE

It's a morning of northern autumn, a day
or so after the election victory of Barack
Obama. The huge multicoloured upheaval
in the country that lies south of us, all that
red and blue, seems as fabulous and remote
as the idea of great embroidered elephants
or whales breaching blue waves.

I think of glimpses of Jesse Jackson's face caught by the cameras while the new black president-elect makes his victory speech, a mixture of pride, abiding joy, even grief: emotions so intense that his face is almost an expression of bafflement. How can one feel so much at once?

I think of my mother, dead a year ago, and how excited she would have been by Obama's victory—how she would have followed the campaign in every morning's paper and every night's television news with indignation and applause, like a member of the audience in the pit at the Globe Theatre watching Shakespeare's company perform *Henry IV*.

I think of the time in my young days when Canada suddenly had a leader who seemed to come out of nowhere with vision and oratory—and how I might have voted as much for the brilliance of Pierre Elliott Trudeau's blue eyes across a packed square as for his policies.

Yet here I am, apparently so far away from it all—from my mother, my past, the great events of my time. We are in a frost interregnum, mild weather after cold stripped the branches. A little moisture overnight; not enough after two months of drought, but grass can do so much with so little and gleams green in the brown morning. I know this feeling welling up in me, even though it is an emotion that is as hard to place as the exact colour of this morning's sky. It is the feeling that, all my life, has made me want to write a poem.

IT'S THE SAME FEELING Robert Frost described when he wrote that a poem "begins as a lump in the throat, a sense of wrong, a homesickness, a lovesickness."[1] But what is this particular tickle in the human brain? Why is it strong enough, urgent enough, that for thousands of years of recorded history—and likely for at least a hundred thousand years before that—human beings have been impelled to find ways of expressing it?

The other day, I came home after the typical round of meetings and errands and personalities, feeling faintly envious of people who only have to run for high office. ("Bet Sarah Palin isn't running stuff to the dry cleaner and picking up cat food.") So I put some music on to cool my overheated brain while I unpacked bags. It was a piece of church music recorded in Belfast's Cathedral of St. Anne. The organ's deep chords growled and then soared, and that same inchoate yearning swept over me.

For the first time in decades of writing poetry, it occurred to me that this response to certain music is exactly the same as the emotion that makes me want to write. It is as though the familiar sensation is as much an *end* of the creative process as a beginning—as though we write poetry or music not just because we are pushed to do so by this emotional blend but also because we want to evoke it in others, to make it resonate in someone else's mind. We actively seek out opportunities to experience it.

For heaven's sake, why? This particular emotion—no, I can't keep calling it "this particular emotion" or it will make for clumsy prose. Let's call it "yearning." Anyway, yearning does not serve an immediately obvious purpose in human evolution and survival. It's not like the well-defined brain pathways that drive us to feel rage, parental love, anxiety, fear, lust. Instead, it's an emotion that seems to exist at the interference fringes where basic drives interact. It also seems to draw on the brain's capacity to remember. It has something to do with simultaneously belonging and not belonging.

Yearning is not the same as aesthetic appreciation, the pleasure we get from recognizing symmetry, pattern and balance. It's not the only reason that people create and absorb art; yearning is not about celebration or rebellion against the status quo. Nor is it the pleasure of craftsmanship, which enters in once you've started to write the poem but is not the triggering impulse. It's evanescent and often impossible to call up at will.

Yet "that Frost feeling" is as distinctive a colour in the mind as red or blue is to our visual systems. You know it when you feel it; you want to feel it again.

Perhaps it's what scientists of evolution would call a spandrel—a by-product that arises as a species moves through the process of natural selection over generations, but that isn't specifically selected for. Yearning might be just a nice-to-have but accidental effect of having brain systems that arose for more basic needs. A good stout capacity for loving your offspring or getting angry when threatened or anxious when you can't figure out what's out there—these are all perfectly sensible reactions to the world and will ensure survival. In a complex system like a brain, there will inevitably be areas where different emotions spill over and interact in relatively consistent ways. That doesn't make such interference states necessary or important in and of themselves.

I'm inclined to think otherwise. Yearning is something that emerges out of our human capacity and need for empathy, our ability to think collectively.

HERE'S MY JUST-SO STORY about empathy:

Mammals learned to make crying noises when infants were separated from mothers—not an issue for fish or turtles, but a matter of life and death for a baby rabbit. Somewhere in the course of human development we added the liquid novelty of tears, for reasons still unclear. Our tears of grief are chemically different from the normal tears that bathe our eyes and keep them functional.[2]

As part of the evolution from primates into early hominids, crying became an activity for the whole community. We retained the infantile expression of fear and discomfort into adulthood; it became a powerful signal to others of a distressed state of mind, and continued to demand that others do something in response—that they find us in our loneliness.

That call-and-response became something new, a kind of resonance; the emotional state of the crying one sets up a similar state in the listener. This is communication. But not communication as it would eventually come to be understood, something that happens in language and is primarily an exchange of information: "The pain is on my leg just above the knee." Instead, it's how a participant in a Catholic Mass is a communicant—an absorbed relationship, an intermingling of states. Somehow, this intimate response to tears has adaptive value for the group as a whole.

Meanwhile...or before...or after...the capacity for language evolves. And with it, the capacity for putting narrative into words. "This happened, then this..." And in the plastic, rapidly developing brain the two become fused—the tendency toward emotional resonance and the capacity to narrate stories. "This happened, then this, and oh, woe!" And all the tribe is weeping.

THE WORD "EMPATHY" only came into our vocabulary in the early twentieth century.[3] It was first used in art theory to describe how someone looking at a work of art projects herself into it imaginatively. It was the translation of the German word *Einfühlung*, the feeling of being one with something outside the self.

The term may be new, but philosophers and political scientists have been debating the same basic idea since Aristotle. At its heart, it concerns a central question about who we are. Are we Rousseau's kindly savages with hearts naturally inclined to care for our fellows? Or are we nasty brutish creatures who need to be kept in line by all the restraints of society? To the general public, twentieth-century science seemed to support the nasty and brutish interpretation. The popular summary of Darwin's theory of evolution—"Nature, red in tooth and claw"—gave us the picture of human beings "progressing" from more primitive ancestors, but only through stiff competition. Driven by such a grab for individual survival, what could make humans sympathize with others and perform altruistically for their good? Surely only the supports of culture and the restraints of government would hold base nature in check.

However, science is starting to show us that empathy is not the wishful domain of poets but a real phenomenon in our brains, and to demonstrate the kinds of mechanisms that produce it. First of all, the same areas of our brain light up when we see someone else experiencing pain or disgust (or love or joy or shame) as when we directly experience the stimulus.[4] It's as if we do not need the actual pinprick in our own finger to travel the same route in our brain.

When you think of it, this emotion contagion is a remarkable resonance. Why would that be a useful thing for an animal to develop? Emotion is costly in terms of energy. Why get excited when I don't have to? I have enough on my plate, thank you.

However, in a species that lives in interdependent kin-related groups, emotional contagion *is* a very practical response. Noticing that another individual is in pain instantly primes your own brain to respond with fear, alarm. You're ready to deal with the tiger in the forest, the snake in the tree, or to watch out for the thorn or the researcher who is going to prick you with a pin.

The second step in empathy is, paradoxically, a distancing. Emotion contagion causes two brains to light up in lock-step. But when you are the observer, a second brain area also lights up nearby—an area that, in effect, says, "Hold on, you are *not* feeling the pain. Don't act as if you were." Then a third set of brain systems, ones that involve judgement and social context, start their trickle-down effect. "It's all right—a woman in a white coat may be sticking a needle into my child, but it's for a vaccination. There, there, Dear. Don't cry."

Once again, there is a deep practicality to such a one-two combination. A complete overlap between you and another isn't what's needed in an emergency. A mother who responds to her child's distress with exactly the same brain circuits as she might use if she were herself experiencing pain won't be able to deal usefully with the situation. Empathy isn't primarily about feeling distress for others but for alleviating it. To do so, it needs a gap, an opening, a difference.

NO FROST HAS FORMED on this morning's soft surfaces. Frost is like empathy; it needs a narrow gap in order to form its arabesques.

On a cold calm evening, when there are no clouds between us and the black ache of space, the earth gives up a torrent of photons carrying infrared radiation along with molecules of water that evaporate from cooling soil. The thin layer of air nearest the ground cannot stop the heat, which passes through it quickly and lingers a little in the higher layers. On such an evening, air is slightly warmer at shoulder height than it is at the soil's surface. Water molecules accumulate in the lowest layer of air, but the cooling air cannot hold as much moisture as it did before and must release them. However, the molecules do not condense into the ordinary flow of liquid. In the stillness, they leap straight from incoherent vapour to crystalline solid to trace a loving outline of surface.

Frost won't form on a breezy night when the wind mixes up the layers of air. Nor will it form when clouds keep the photons of thermal radiation bouncing back and forth rather than passing undisturbed out to space. Only when there is calm, a space, a difference, can frost write its elaborate adhesive lines.

THE BRAIN CIRCUITRY of empathy is central to our response to art. Researchers can do their experiments without actually sticking pins into a real subject. We also respond empathetically when we are given a representation of the stimulus—a picture of a foot in a painful situation, a descriptive narrative.[5] From this springs our powerful capacity to engage with story (and perhaps our rooted pleasure in art that represents our world in a recognizable way).

Once again, our ability to respond to words or images as we do to real events seems an odd capacity to develop. Why get ourselves all worked up about something that's not really happening? Until you think of the practical advantages. Feelings are largely about anticipating the future—an evolutionary strategy mammals have been working on for hundreds of millions of years. There's a great advantage in being able to assess other entities in your environment and figure out what their emotions are toward you. Our earliest ancestors learned to do this kind of forecasting by observing the actions of others and running them through their own brain maps: "When *I'm* making that noise, I'm feeling mad as hell. So that guy over there is mad as hell. I'd better back off."

Rats and cats and elephants all are capable of this kind of calculation—without, of course, that little mental commentary because language isn't available to shape it. But the way they carry it out is essentially the same as the way humans do: by having the same brain areas light up as are firing in the individual they are observing.

When language did develop, humans had another tool with which to assess, practice and share the feelings of others. After all, you can't wait for a crisis of pain to begin figuring out how to deal with it, how to respond appropriately. Practicing in advance helps prepare. So we seek out opportunities to experience the sensations of empathy, and we respond to artifacts—sounds, stories, images—that create such opportunities. And we have to start this training early. We teach/are taught from our earliest moments of breathing air that we are part of a community of communicants.

The anthropologist and scholar Ellen Dissanayake has proposed that the fundamental processes involved in making art are rooted in the parental bond, in the instinctive things humans do (regardless of their particular culture) to interact with babies.[6] We exaggerate our expressions, our mouth movements. We repeat, repeat, repeat. We start with basic patterns and elaborate them. We make routine things special.

We enter into a rhythmic responsive dialogue. Then we carry this pleasure in exaggeration, repetition, elaboration and "making special" forward into a whole new set of activities, activities that bond adults into the rhythmic responsive relationship we learned to enjoy in our earliest period of life—the empathetic relationship.

Empathy is a remarkably elastic capacity. We can feel it for virtually anything, even a different species. When I was a little girl, I used to tuck pieces of blown paper under bushes, not because I was a particularly tidy child, but because I wanted things to be safe, looked after. I felt as if I were being blown along the road myself and wanted to make a scrap of paper feel loved.

Sounds rather sweet. But empathy can also be as dangerous as any other of the great emotions, largely because it requires maintaining a delicate balance between closeness and distance. It's all too easy for those top-down judgement processes in human cognition to rule that empathy doesn't apply to "distant" individuals, those wearing turbans or burqas. The sense of deep understanding possible in a small tribe, whose experiences and language are much the same, does not necessarily extrapolate to a diffuse and diverse humanity speaking the world's many languages. Empathy may afford us an illusion of understanding more than we actually do.

There's also the danger of feeling that we've shared in the weeping and have therefore resolved the situation, because that's what emotional sharing does when it happens in a small group. It binds individuals; it reconnects the lost and lonely ones. But crying over the horrors on the television news is a one-way street. It achieves no such connection because it has moved from one kind of story to another. It goes from the real and unique story of events—"this happened and oh, woe!" to *a* story— the fictionalized narrative that we learned to enjoy because it seemed to give us that sense of emotional connection, even as we recognized with pleasure that it *wasn't* real. An emotional response to an event plugs us into the neural systems for empathy, but that empathetic display also disconnects us from reality. We don't have to do anything in response.

BUT ON THIS AUTUMN MORNING that feels so far away from everything, I do not have the urge to tell a story. I am in that state that reaches for a poem. So let's go back to that Frost feeling. What is it?

Yearning is, at its heart, a simultaneous sense of belonging / not belonging. It's linked to memory, a sense of being displaced in time, of indefinable loss—yet a loss that is somehow repairable by action, by expression, articulation.

It can be triggered by sensory experience, often something in the natural world like a dawn sky or a line of hill. However, yearning is not simply the perception of colour or line. It is also the feeling of being connected somehow to it. Yearning can also be triggered by human works—the sound of distant music (bagpipes on hills if you were raised in Scotland as I was) or certain kinds of space like cathedrals or quiet rooms. It's not something you are likely to feel in a crowd. Although you may feel intensely lonely and disconnected in a shopping mall, it's not this sense of yearning, which seems to need some space around it.

It is a tangle of emotions. It has some element of the tenderness associated with our deep response to small children—an urge to protect, hold—but it is the desire to hold a moment, a pattern in

your mind, rather than a furry toy. And yearning can parti-
cipate in a kind of wild joy at the beauty of the world; it can be
painful. However, it does not include anger or rage, nor is it
fearful. And the emotions tangled in yearning's web are
somewhat muted, untargeted, as though they are being held
at a distance.

My hypothesis is that this sensation of feeling connected
and simultaneously separate, coupled with an urge to resolve
the tension by expressing it, draws on the same brain systems
as empathy does. I've no idea how to design the research
experiment, but I'm willing to bet that when I am experiencing
that pre-poem Frost feeling, brain circuits overlapping with
empathy would light up in the MRI images. This would explain
the curious persistence of this apparently marginal sensation
in human experience, and why we draw on it to produce
poems. We are practicing something essential.

> *Poem I wrote in Quebec City at 0330 27 Oct 09. I woke up,*
> *couldn't sleep (approx 0830 Cyprus time) and anxious to fly*
> *home that morning. I just started writing, first time for me.*
> *I wanted to share this with other Military.*

THIS NOTE CAME with a submission to a project I was involved
in recently, one that worked with members of the Canadian
military and their families to create poetry out of their experience. Many submissions came with similar sentiments—the
first poem I ever wrote, thanks for the chance to share it.

Poetry has a special connection with yearning that is even
stronger than other art forms. After all, what is the classic
teenage response to angst? To write a poem. To say in effect,

"I need to say this, make these sounds; I hope it will evoke a response in you (even if it is kept secretly in a wallet and never shown to the object of desire or the agent of pain)." So many people have written the only poem they ever will write in such a state of yearning. They carry it around, keep it in a notebook or an envelope for years or decades. They seek out the music that makes them feel this way, lie on their pillows and then assemble words that will assuage the emotion yet make others experience it, too.

The urge to assemble words into a poem comes because empathy enables us to inhabit simultaneously the first person and the third person—to experience the pain in the foot directly and also at a remove. It allows us to move our brains across that narrow gap.

The two-stage feature of empathy, I believe, underlies the therapeutic power of writing. It's not simply that you feel better because you've "expressed" emotion. The first-person outpourings in our journals or into a friend's ears are not where healing truly takes place. The assumed value of "expressing yourself" is based on Freud's old model of the brain as a hydraulic system: contents under pressure; open the safety valve. In fact, therapy lies not in releasing emotion but in *absorbing* it and doing something useful in response. Essentially this process moves you to that hybrid space where you inhabit both first- and third-person states at the same time, where you can empathize with yourself.

All arts have a therapeutic quality, of course. The physical slap of clay, the focused mixing of colour, move us away from immediate pain and give us something else to concentrate on. But because writing is done with words, the effect is more intense. You are working directly on the subject matter of your own narrative.

Even the basic outpouring, the organizing of sentences out of selected words, will begin the process of shaping. But it's only when you take those words and make something more structured—find a rhyme, an image, a beat—that the empathetic process kicks in. So it is no wonder that teens turn to poems when they are in love or in heartbreak and clutch the verses they have made. Such poems are evidence they can empathize with the person within them who went through the intense experience.

We tend to dismiss this as the province of "amateur" poetry. We might grudgingly admit this kind of poetry writing might be useful as a stage to pass through, but we think it should be left behind ASAP. "Professional" poets of our time are so dreadfully afraid of the amateur, the muddle of cliché at the open mike. It's as though we're think it's catching, that it creates some awful average that will suck us into its maw, that audiences will become so accustomed to bad poetry they won't know the good stuff. But part of our pleasure in poetry is similar to our pleasure in watching really good athletes—a young man streaking toward the net in a hockey game, a young woman racing down a ski slope. It's because we have a shared body plan. We know what it feels like to move. We have a basis for imagining what it might feel like to have that strength, that agility, even if we could never do these things ourselves.

Poetry is also something we know from the inside—what it's like to speak, to assemble words, to try to use them to express an emotion, a thought. So we take the same kind of pleasure in witnessing it being done well. And we know when it's done only at the basic level that anyone can master. Our capacity for mapping the actions of other bodies in our own brains ultimately gives us a basis for assessing how well another has succeeded. Because we are designed for empathy, we have a basis for assessing differences.

I HAVE FINALLY REALIZED that the Frost feeling lies at the heart of Wordsworth's definition of poetry as "emotion recollected in tranquility."[7] For years I disliked his phrase, which made poems seem like limp and elderly cousins of passion, mere faded reflections. But Wordsworth was quite right. However, the gap between emotion and recollection is not the wide span of years but the narrow cleft in a human being between "me" and "her."

So here I am this morning, wanting to write you a poem. "You" the reader, "you" the writer, both of us tangled up together in the same brain. The sky is soft as feathers. Where do we go from here?

TWO Metaphor at Play

I'm rinsing out the bathroom sink. At my
feet, Pushkin is poking a toy mouse under the
blue plush of the bathmat then pouncing on
it where it is hidden. He scoops it out of its
soft cave, tosses it in the air a few times, then
pokes it back into the hole. Then he lurks
with his chin on the floor, his tail twitching,
until he pounces again and the toy mouse is
sent spinning into the air.

It suddenly occurs to me that he is *pretending* it's a real mouse. He's not under any delusions that it *is* a real one—he's not fooled into trying to eat its straw stuffing. He is imitating the instinctive actions that a cat uses to hunt, but the activity is quite voluntary. He's having a great time.

For the first time, I realize that play is a process of metaphor. It acts "as if." It draws on an ability to hold two situations in mind at once—a real world and a pretend one—and to fool around with the combo.

WE HAVE TRADITIONALLY THOUGHT of metaphor as arriving in the human brain as a late-blooming ornamental facility that needs deliberate thought and practice. I remember grade-school exercises where we were set to laboriously pick out metaphors or similes in a piece of writing. Then the grade-seven teacher gave us the assignment to "make up a metaphor of your own." I remember looking at the other puzzled faces in the classroom. It was as if she had asked us to twitch our ears. Maybe there are muscles for doing that, but you've never located them.

Later, in university, I was given ever more advanced words to shave the idea of metaphor into smaller slices. *Metonymy*, for that habit we have of using one thing to stand for something related. ("She lives by her pen" does not mean that she chews the barrel for sustenance.) *Synecdoche* lets a part stand for the whole (as in the phrase "blue-rinse set" to describe elderly ladies). I was impatient with such distinctions, thinking them artificial, thinking they mask something more fundamental—a process of laying one thing over top of another and seeing where they are the same, where they are different. I never did learn how to pronounce "synecdoche."

In fact, the underlying process *is* fundamental to all of language— and indeed, to how we think. As Mark Turner points out in *The*

Literary Mind, our thinking in language is based on what he calls "small spatial stories."[1] We learn these consistent, minimalist combinations of events from the time our newborn brains are starting to sort the rush of incoming sensation into something useful.

Objects can drop through the air or roll across the floor. Liquids pour from one bounded container into another—or spread all over the table. Some objects move in ways that indicate they have intentions toward us. Our brains piece these sequences together from bewildering incoming data—the ball flies through the air, and at every microsecond we receive different sense impressions of shape, speed, sound. But we bundle them together into a single event narrative and know they belong together, that the thump against the floor means something has happened in the immediate past and will continue to rebound into the immediate future.

These small stories about how objects and actors move become so ordinary a part of our experience that we no longer pay attention to them. But they shape language because we are continually taking them as a template for other experiences. Something follows a path toward a goal, is blocked along the way, flows around it. That template can apply to snow melting or the plot of the hero myth.

Turner points out the tremendous agility of our brains in applying one story over top of another. Take a proverb like "when the cat's away, the mice will play." Proverbs are highly condensed versions of narratives that we unpack with dizzying speed. We can apply "the cat's away" to situations that range from a rowdy classroom to a banking system with insufficient regulation. Not a cat or a mouse in sight, but we make the metaphorical pattern fit effortlessly.

Douglas Hofstadter maintains that the making of analogies is not just a separate function of the brain like recognizing shapes or making syllogisms, but that it *is* thinking.[2] That, in the constant back-and-forth between memory and present sensation, we lay down a kind of grid—one that gets larger and more finely meshed all the time. We lay this mental grid over whatever we have to deal with at the moment and compare "now" with what we have stored, perhaps in the same way a painter might lay a grid over a blank canvas in order to reproduce a painting.

He suggests that this process of analogy is what we enact whenever we choose a word or respond with a phrase to a situation. Essentially we do a lightning-fast comparison. To me, this makes a lot of sense. And it explains why poetry is so central, why we feel it is an important discipline even if few people pay much attention to it. The process of making analogies—metaphors—is central to our mental functioning, and it's also central to poetry.

METAPHOR SURVIVES even when language itself is fragmenting. "Look at the frogs," my father said. We were sitting in a tired coffee shop in a downtown office building, while my mother finished a medical appointment upstairs. Dad, bundled in a puffy brown parka and wearing his dark-green tartan "bunnit," was staring out the window at the street.

Startled, I turned around. Frogs on a concrete sidewalk in a prairie city? At this time of year? Outside, I saw curled brown leaves were being whipped up by a wind around the building's sharp corner and skipping up and down. They didn't look anything like frogs, except for their unusual motion.

By now, language was a system in tatters for my father, who had been supremely quick and inventive with words. He could no longer articulate the stories he had hung on to fiercely for as long as he possibly could—the narrative of his young life, the incidents that had shaped him. We had been driven almost to a hopping frenzy by having to listen (and respond) again and again to the same anecdotes.

All that was left were formulaic phrases. *Pleased to meet you. Hello. I love you. Look at...* He could no longer shuffle words into novel arrangements. But even so, he could observe and reach into the scrap-bag of his mind and put a motion together with a memory.

He said it slowly. "Look at the...frogs." The image was there in his mind but he couldn't get to the language. When I looked at him, puzzled, he flapped his hand up and down, smiling, and gestured to the window behind me.

My father had always been an imaginative, playful man. His old Volkswagen van, with *Bill Major Painting and Decorating* in shiny peel-and-stick letters on the door, became the starship *Enterprise*. Toronto Transit buses became the enemy Klingon ships, and he and my little brother got so excited by one game that Dad swung the van the wrong way down a one-way street to get away from a Klingon attack.

When Alzheimer's began to shred his brain, I noticed that he tended to see faces everywhere—in lilac branches outside the windows, clouds, figures in the carpet. The line of the mountains as we drove toward the Rockies traced profiles for him, faces turned up to the sky. Sometimes his tendency to see faces became hallucination. "Who are those people over there?" he would ask, and get up to look more closely at a group of ornaments on a bookshelf.

It was as though he had returned in part to that state of an infant looking at the streaming array of incoming sensation to find the configurations that look like a human face—the most significant feature of our young lives. The ability to map what *looks like* onto what's *really there* and assess similarities and differences is so central to our lives, from beginning to end.

AS A POET, I am constantly trying to twitch those mysterious muscles with which we make up metaphors. Poetry has always seemed to have an essential relationship with figurative language— all the tropes from anthimeria to zeugma.[3] So it's no surprise when researchers find that lyric poetry really does have a larger percentage of active and extended metaphors than do other kinds of writing.[4] "If it weren't for our ability to compare one thing to another, then to draw seemingly spontaneous knowledge from the comparison, poetry would be impossible," writes poet Stephen Dobyns.[5]

In my experience, metaphor is built in at the most formative stages of a poem. Once you've decided that you are moved to write about the dawn (or love, or your father's dementia), there begins the process of sifting through your brain to find some template, some parable or connecting hinge. Anything that will take the relatively simple and all-too-specific experience you want to describe and open it out, give it resonance.

There's something almost visual about this process; it is as if you were holding two images printed on celluloid, one on top of the other, to look at the light coming through both at the same time. Having written fiction, as well, I find the process of invention to be different in the two genres. Of course, there is also a visual component to imagining fiction. For me, it's as though I'm watching a film—the actors are positioned here, the setting includes that vista and these objects, the camera is shooting from this angle, the character will move forward and say...what? But this is a relatively uncomplicated process of visualization, as though you are only looking at one scene at a time. It doesn't quite have the layered tangle of the poem, the insistent sense that two (or more) things have been superimposed.

The poem might pull together the pale apricot colour, soft as feathers, of the sky in the east with a phrase free-floating in memory, "the bird of dawning." Which sends me to the source of that phrase in the first scene of *Hamlet*, where soldiers are huddled on a pre-dawn battlement, wondering at the ghost of a dead king and feeling the times are out of political joint. The bird of dawning referred to in that scene is the cock, said to cry all night at the time of the Saviour's birth to drive away ghosts and evil spirits, "so hallowed and so gracious is that time."

The poem will grow, not as a neat correlation of $x = y$ but as a tangle of connections between real birds, the idea of political ferment, the cold weather. That scene from *Hamlet* does not create a map that insists all points in my new poem must correspond to some feature of the original. Instead the original is a kind of web wherein relationships between its nodes are similar to relationships in what I am trying to describe.

MY POETRY BOOKS are full of little pencil marks beside metaphors that have stopped me in my reading tracks. Take Derek Walcott's image for a chainsaw attacking a tree trunk:

> ...*The generator*
> *began with a whine and a shark, with sidewise jaw,*
> *sent the chips flying like mackerel over water.*[6]

My neighbour revs up his chainsaw to cut down a sprawling Manitoba maple that's in the way of a new garage, and I see that shark, the murderous teeth, the impossibility of escape. The picture maps onto my own experience as quickly as it maps onto the description of a gommier tree being logged in the distant Caribbean.

Or a poem by Rhona McAdam, in which her mother's failing memory becomes a spinning wheel that "spins backwards, then not at all."[7] I think instantly of my father, the way his stories slipped backward in time, swung there for a time, then ceased. Or that moment in Dante's *Purgatorio*, when Dante the Pilgrim, feeling sleepy, is overtaken by "a thick rush of souls"— the shades of those who have been negligent and slothful while alive. They cry out:

> *Faster! Faster, we have no time to waste,*
> *for time is love.*[8]

I reread Dante around the time that I was caring for my parents near the end of their lives. This particular image reached out to grab me by the throat, for, of course, time is love. I was spending so much of my time at a gallop, trying to hold things together. *Faster, faster,* was the litany of my day, as their lives became suddenly smaller and slower. It was like living in one of Einstein's thought experiments about relative motion. But, cranky, disoriented and dishevelled, I still knew that the time I was expending was indeed love.

FANCY "POETIC" METAPHORS like the ones dreamed up by Walcott and Dante may be arresting, but they are only an efflorescence of a process that goes on constantly in our brains. Cognitive scientists have come to realize that meta-phor is central to *all* language. Even something as small and apparently transparent as a preposition is a metaphor as often as not. The cat food may be literally *in* a plastic tub *in* a real kitchen cupboard. But a specialist in feline medicine is not *in* any kind of bottle, jug or closet. Instead, we have visualized an academic discipline as a container.

"Your English prepositions," moans a Brazilian writer friend who has been struggling with them for years.

But prepositions get you in any language that has them. In English, all sorts of concepts are viewed as relationships to containers—we work *in* physics, our books are *in* press, we are *in* the middle class. But even in a language as closely related to ours as French, underlying actions are often visualized differently. People are *of* a class, as if the class owned the individual, rather than being a box that the individual could jump in and out of (or be constrained by). Books are *under* (*sous*) press, as if they were literally being pressed down.

Learning language is learning an invisible web of metaphor. Take a verb as ubiquitous as "take." It's rooted in a simple action: to move a physical object from its original holder, with a suggestion of overcoming at least some mild resistance in the process. But in the phrase "take the expression" we've moved away from the physical concept and are talking about the process of moving an idea around. We can *take* buses, courses, lovers or conniption fits without noticing any metaphorical underpinnings.

Metaphor theorists like George Lakoff call these familiar relationships "conceptual metaphors" and put them in capital letters: AN IDEA IS AN OBJECT (which allows us to say ordinary things like "I take your point" or "she grasped the subject") or EVENTS IN TIME ARE EVENTS IN SPACE (which allows us to say "I'll feed the cat *at* five o'clock" as though the time on the clock was a position).

"What about CATS ARE HUNGRY ANIMALS?" Pushkin interrupts in his loudest voice.

"No," I tell him. "That's a literal statement, not a metaphor. And cognitive scientists are not yelling. They're just trying to make the generalized underlying patterns apparent. And we haven't arrived at five o'clock yet."

"Oh, well..." He stalks back to poke the bathmat again.

Some conceptual metaphors appear in virtually all human languages, indicating they reflect some basic underpinnings of our bodily experience as patterned in the mind. For instance, the metaphors of ANGER IS HEAT or ANGER IS PRESSURE are remarkably consistent couplings across cultures.[9] Others reflect connections that specific cultures are inclined to make, though they can still be surprisingly widespread. For example, languages as different as English and Chinese develop metaphors related to "face" in surprisingly similar ways.[10] Both share a pattern that progresses through ideas such as a face representing felt emotions ("he was long-faced"), to being the outer appearance of something ("the village wore a placid face"), to being equated with dignity or prestige ("he didn't want to lose face").

Words from "spirit" to "Internet" all got here the same way, by a process of observing and imagining how things are like other things. I'm particularly fond of the derivation of the word "muscle." It comes from the Latin *mus* (mouse). The ancients saw the twitch of muscle under skin and thought it looked rather like mice moving there.

(Pushkin would quite understand this connection— it's exactly like the way toes move under the bed sheets, something he loves to leap on. "*Pushkin!*" we yowl and throw him off the bed.)

I'm enormously pleased by the idea that two words as different in meaning as "muscular" and "mousy" could emerge from our sensory experience with small rodents.

"Nope, no mouse here," says Pushkin at last, leaving the bathmat crumpled on the floor. "Should we move on?"

METAPHOR MAY BUILD LANGUAGE, but the capacity to create metaphor is, in a sense, *pre*-language. It is built on a platform, a capacity for comparison, that evolved long before we were capable of any verbal performance more complicated than hoots.

Neuroscientist Antonio Damasio theorizes that "core" consciousness resides in a comparison of "in here" and "out there"—an ability we share with other species. We are not particularly conscious of our bodies, the state of our viscera, our balance, the regulating system that keeps our temperature within a narrow range, the mechanics of our breath, the chemicals being released into our brains by somatosensory signals. But our bodies are sending a continuous stream of information to the brain, which becomes a stable framework of reference for assessing what's coming from beyond the body's boundary.

This process of mapping inner against outer began with the earliest organisms—as Damasio points out, it takes something like perception for a single-celled organism to sense the state of the chemical profile within its boundary and an "unconscious knowledge" in order to respond to it. The continual evolution of an ability to perceive and respond leads slowly to systems like vision and complex emotional reaction.

"It's intriguing to think that the consistency of the internal milieu is essential to maintain life *and* that it might be a blueprint and anchor for what eventually became a sense of self in the mind," writes Damasio.[11]

Core consciousness comes before language. We share it with amnesiacs who have lost all sense of their personal narrative, with dementia patients who have lost language almost entirely. We share it with other animals—the cat poking at a toy mouse, the real mouse in its nest of dry grass.

No animal experiences the world through a single modality. We touch/see/hear things at the same time and associate the sensations firmly together. This facility for associating information of different kinds is essential to the brain's functioning at all levels and lays the groundwork for metaphorical speech.

JUST WHAT DOES HAPPEN in the brain when we handle metaphor? There has been a burst of research into this aspect of language over recent decades, as new technologies for watching the brain in action have become available. They tell us, for example, that we recognize a metaphorical connection as fast as we recognize a face or a joke. Like so many of the apparently simple things we do (recognizing a voice, picking up a small object), identifying a metaphor requires enormously complex computation. People trying to build artificial intelligence systems to handle interactive speech or translation have been stymied by metaphor.

From Aristotle on, we have tended to think that the brain must handle figurative speech differently than it manages literal speech. This idea seemed to gain some support in the 1970s from brain-imaging studies indicating that the right hemisphere becomes engaged when the brain is required to handle metaphors. Cognitive scientists hypothesized that the normal pattern was for the brain to "look up" a literal meaning for any words presented to it and, if no match was found at that level, to hand this fragment of speech off to the right hemisphere for assessment as a metaphor.

However, more recent studies have largely discounted this view. Very detailed timing studies of how long it takes to process speech indicate there is no difference in the time required to handle literal speech and familiar metaphors.[12] Researchers with electrodes can pick out the onset of wave-forms that indicate when we have integrated the meaning of words in a sentence—a process that typically takes between two hundred and four hundred milliseconds. They show we understand a familiar image like "My aunt is a battle-axe" as quickly as we understand "My aunt has a bad temper." In other words, in real-life studies there is no time for our brains to look up "battle-axe," choose "No, not a real battle-axe" and hand the word over to another assessment process.

What *does* get handed over to the right hemisphere is anything that gets bogged down in the lickety-split language look-up system centred in the left hemisphere. This happens most frequently with novel "literary" metaphors, but also if there are surprising contexts for a literal meaning. The right hemisphere casts a wide net of inference and memory to examine potentially looser connections in order to find and assign some meaning.

So metaphor isn't special speech. But it does have some particular tendencies. First, we use metaphors a lot when we talk about emotions. You can easily talk about buying groceries in literal language: "I got a shopping cart but it had a stiff wheel, so I took another one. I needed milk and bread." But it's almost impossible to talk about your friend's love life without metaphor: "She really *fell* for this guy. But then he *dumped* her." This is unsurprising. Emotions aren't visible objects to pick up in the grocery aisle, so we are forced to reach for other ways to name and describe them. But this is probably another reason that metaphor is found so frequently in lyric poetry—a genre that is used so much to express how we feel.

Another feature I find intriguing is that verbs seem to lend themselves to metaphor more easily than do nouns. Studies show that, in English, verbs are more frequently used as metaphors in ordinary speech, and that we notice noun metaphors more easily.[13] Once again, this is unsurprising. Verbs express relationships, interactions and transitions, so they express metaphorical relationships and interactions rather naturally. Most nouns have a certain intractable stuck-to-the-world quality, which tends to make it surprising when we apply them to something else.

It's this quality that makes Walcott's chainsaw-shark metaphor more arresting than the more subtle metaphor of the chips "flying" through the air. Flying is something that, in a literal sense, requires a purposeful flier. Birds fly on their own, as do trapeze artists and squirrels. Planes fly as piloted mechanisms. But when leaves "fly" in autumn, a quiet figurative transition has taken place—they have become personified as purposeful agents. "Let the chips fly" is a cliché, based on the same process of animation used for leaves. But Walcott has turned the inanimate wood chips into mackerel flying briefly above the waves, paralleling the transformation of the inanimate machine into an animate purposeful shark. The verb is just as much a metaphor as the noun is, but it's completely overshadowed in our attention by the comparison involving nouns.

Finally, I'm interested by findings that in ordinary speech we often tend to use familiar metaphors as a kind of social lubricant—a way of marking transitions in subject or wrapping up subjects in an inoffensive way. "Well, there you go," we say and sigh to end the conversation about our unhappy friend, even though no one's actually going anywhere. Metaphor is not used always to surprise but sometimes to reassure.

This leads us past cliché to what cognitive scientists refer to as "dead" metaphors, connections that have become so thoroughly cemented that it may seem any figurative aspect has been lost. In such cases the word has simply taken on a new literal meaning, as "muscle" did when the mouse connection disappeared entirely from view. Some theorists posit this as the end of a "career of metaphor,"[14] a kind of conveyor belt that carries linguistic constructions from novelty to novocaine.

But metaphors are never as dead as they might seem. There's evidence they do not become simple lexical items but remain alive in the brain. Some of this evidence comes from studies of the gestures that human beings use naturally and almost unconsciously to accompany speech.[15] Careful analysis indicates that we're very likely to echo the almost invisible metaphor with the physical pattern—for instance, someone talking about "pushing the limits" may literally push into one palm with the other hand.

And clichés have a vampire-like ability to resurrect themselves through hyperbole and pun. Advertising copy constantly appeals to this universal faculty of recognizing the metaphors underlying the familiar. When a realty company advertises its service as being "above the rest" with a hot-air balloon for a logo, they are playing on the idea of physical height that is almost invisible in phrases like "upmarket" or "high quality."

IN MY EXPERIENCE, there are three main equations that underlie most of the literary metaphors I make up.

First, there's the direct comparison of one sensory field to another. One of the most striking examples I've ever come across occurs in a poem that, to my embarrassment, I can't remember the title or author of. It describes a man fixing a drink for a woman, someone he hopes to seduce. The nippled end of the lemon needed for the cocktail is superimposed on the image of a woman's breast, a visual connection that's direct and easy to make. When the poet presents the second stage of the image—the knife slicing the end of the lemon off—you wince because you instantly transfer the sensory details of blade, slice, sever from fruit to human tissue.

The second process is to compare a sensory field to some abstract concept. Emily Dickinson's "Hope is that thing with feathers" is a familiar example. Hope has no fixed sensory qualities of its own so the ones you choose for it can be elastic as long as their emotional resonance is right. The feathers and flight of birds works nicely, though we're more likely to think of hope as robins returning in spring than as a carrion crow. (Although, in context, a vulture circling a battlefield *could* be a macabre figure of hope. The process is definitely elastic.)

The third comparison process is abstract to abstract, such as Dante's "time is love." In a sense this is the easiest, yet it's still constrained by the physical world. Dante's slothful penitents running around the mountain of purgatory to re-learn urgency are embodying time *and* love in a kind of triangulation back to real sensory fields connected with a human being in haste. So in this kind of metaphor you have to find such a real-world counterpart for two concepts at once, making it a little harder to do.

Of the three processes, I find the first to be the hardest. The physical world is very demanding. It looks and smells and feels a certain way and you depend on that familiarity to make comparisons that work for a reader. You can't say "a robin is an oyster" and meet with anything but blank faces. Of course, our metaphorical brains are so elastic that you can always force a connection somehow—"the robin is an oyster that opens on the pearl of spring song." But this is really another triangulation back to a more abstract idea of spring that is providing a common denominator for the yoked-together concepts.

The chainsaw/shark or lemon/breast correspondences don't need those abstractions. They make one-to-one maps of sensory experiences, and we decide whether or not they fit. But even though the world is so full of similar patterns, it's not always possible to come up with a really new way of comparing two sensory experiences. Most of the obvious sensory metaphors have already found themselves incorporated into day-to-day language. Take the sound of rain—it drums, it scutters like small paws, it lashes or beats. Just try and think of a word to describe that sound that hasn't become familiar already.

The inherent fuzziness of an abstract concept makes it easier. It may also be that we can go to different parts of our brains to help with such comparisons. Concrete nouns activate widespread parts of our sensory cortex, while abstract nouns are focused more exclusively in a particular part of the left hemisphere.[16] So it may be easier to open up sensory connections that haven't been used before for abstract nouns. Justice can be—well, just look around the room. A table. A carpet. An overhead light arbitrarily flicked on and off. Or poetry can resemble

> *a hundred and fifty cats' eyes*
> *pickled in vinegar*
> *to see immortality.*[17]

"I don't care much for that one," says Pushkin.

THE POETRY-AS-PICKLED-CATS'-EYES image is from a poem by Miroslav Holub, translated from his native Czech. I particularly enjoy his work because he was an immunologist and his work as a scientist informed his poetry. I have no idea what the sounds and sensory patterns of his original poems would be, but his arresting images *are* translatable.

It may seem paradoxical that one of the most distinctive features of a language-based art is *not* created by the brain's busy words. But to me this makes perfect sense. As Stephen Dobyns writes, "It is the ability of metaphor to elicit large, nonverbal perceptions that is one of the great strengths of poetry."[18]

DOBYNS ALSO HAS A METAPHOR for metaphors. "Simply stated, every metaphor is a riddle, since...the reader always asks how A is like B."

I think metaphors are less like riddles than equations—a metaphor is the equal sign between two apparently different things. It's a symbol that says "these are the same" and also the sign of an action, of a potential. $E = mc^2$ doesn't mean simply that energy is the same as mass multiplied by the speed of light squared. It means that energy and mass can be transformed into each other, that one mode of being can become another.

The range of different types of equation used by mathematicians seems a lot like the continuum of metaphorical relationships we handle in language. At one end is the process of mapping x on itself. The equation $x = x$ may seem trivial and unproductive, but this identity is essential. You couldn't do arithmetic without it. In language, it's like the situation when we use words literally, according to their commonly understood meanings. When you say, "Pass the salt," you don't expect to be driven along a seashore.

This basic identity may make routine conversation possible, but it doesn't create new connections. Mathematicians move quickly on to linear equations like $x = y$. This draws a nice straight line on a graph at a stable and predictable angle. It's like a metaphor that has become a cliché. He's an "old salt," we might say of an older man who has been some kind of sailor. This expression assembles a fixed constellation of images— a stereotype. Weather-beaten face, an ability with knots, an adventurous past and a diminishing present. Probably a skipper-type hat.

Linear equations are not always so dull. Make that equation $x = 3y$ and your straight, 45-degree-angle line pitches to slope steeply upward. Change the numbers again and the lines tilt back and forth like the masts of a fleet of tall ships in a gale.

Poets manipulate cliché all the time to take advantage of these stereotyped connections and make them interesting. Let's make the "old salt" a cat. After all, it seems that orange cats have a sailor connection—the genes for orange-and-white cats are found more frequently in coastal areas of Europe where the Vikings visited.[19] So let's imagine an orange cat snoozing on the dock of a Newfoundland outport, surrounded by the smell of salt and fish heads. He is resting after his long lives, the voyages that have taken cats of his colour to Scandinavia and then to Ireland and then across the miauling Atlantic.

"Don't be silly," says Pushkin. "We're out here on the prairies."

"I'm just *playing*," I tell him.

"Doesn't look like much fun to me," he replies. "And what about 'Write what you know'? You've never even been in a Newfoundland outport."

"We know all sorts of things we haven't experienced," I tell him, annoyed. "And like Stephen Dobyns says, we know them because we can think metaphorically."

But he has lost interest in epistemology.

MATHEMATICIANS AND POETS need more than straight lines to describe this world's lovely muddle. One of the first things they do is to intensify its relationships. Multiply x by itself once, twice, any number of times, and you get polynomial equations that draw curving lines.

The parabola's stately sweep is drawn from the equation $y^2 = ax$. It sometimes seems to me that basic quadratic equations like this are a lot like the big metaphors that shape our brains, the narrative lines like "spring is like youth" or "the cosmic egg." We take the idea of an individual human experience of youth and intensify it into the growth of the year. We take the idea of watching a chick hatch from an egg and multiply it by itself millions of times, until it becomes the idea of the birth of the world. These basic metaphors generate a satisfying combination of simplicity and infinite scale, a smooth curve that arcs across different human societies.

But things get *really* interesting in mathematics when you start to multiply the terms of an equation by the square root of minus-one—an imaginary number that you can't get to from here in the normal, step-after-step sequence of the number line. One times one gives you plus-one, but so does minus-one times minus-one. There is no number on that line that you can multiply by itself to get to minus-one.

Undeterred, mathematicians have said, in effect, "Let's *pretend* there is some number that's the square root of minus-one." From this playful leap, they take mathematics to another plane—the complex number plane—because you can multiply any regular number by this imaginary quantity called *i*. This gives you a whole new number dimension that turns out to be significantly connected to how our world works. For example, the laws that govern the universe at its tiniest scales are governed by the complex number system.[20]

Complex numbers make for interesting equations. Their solutions aren't lines at all in the usual sense. The Mandelbrot set, for instance, lives in the complex number plane. It's that bubble-birthing-a-bubble shape you see illustrating books about fractals, and it redefines the idea of "border." Lower-dimensional equations draw nice neat divisions between what's in a set and what's not. Zoom in on the border of the Mandelbrot set and you will see constant irreducible ferment. It never settles down and simplifies. You could go down to a million decimal places and the line is as frothy and curlicued as ever—a border as complicated as the line between salt sea and sand. In the foam froth and hiss of withdrawing wave, does this grain belong to the world of land or the world of water?

This is the kind of world in which poets create new metaphor. Is a hawk like a handsaw? Certainly not. Yet, look a little closer, and it might be—as a shark can be like a chainsaw when you examine the possible connections.

Maybe I could drive this metaphor about metaphors even further. There are differential equations and quaternions and...

But I'd better stop. My poems have sometimes been critiqued for overusing metaphors until they spin like wheels on an icy road. Whoops, there I go again. Pass the salt.

I LOVE THE FACT that mathematicians use the symbol *i* for the square root of minus-one. It reminds me of the fashion for dropping capitalization from poems, so that the pronoun "I" became "i"—a slightly different individual.

AN ENTRY from my journal, ten years ago:

> I was just reading a little monograph by Ivan Illich, "H_2O and the Waters of Forgetfulness." In one of the footnotes, he comments "we lack an independent classification of smells analogous to that of taste (sweet, bitter, salty), as Aristotle has already noted in <u>De Anima</u> (2.9). We cannot indicate to one another what we smell, except by analogy with another sense or by indicating the object that we smell."

I've read similar comments before about the inadequacy of our language to express smell compared with, for instance, colour. It seems to me to be analogous to the comments about Inuit languages being "richer" in a vocabulary for snow or Gaelic not having a word for "no."

We do have general categories of smell—resinous, flowery, putrid. However, they do not fall along a continuum as colours do. Our perception of colour reflects the experience laid down for us by light's spectrum. There are more categories of smell and they are less easy to organize. But they're still quite recognizable to us.

But I think the difference in perceived ability of English to express smell is due to a less complete process of metaphor. There is no mode of sensory perception that does not depend on analogy or reference to an object to evoke descriptive recognition. But in colour, for instance, we have become so accustomed to "jade" or "amber" or "peach" as descriptive of bath towel shades that we forget the phrasing was originally something like "a green like jade." Puce was the colour of a flea. Cerulean the colour of the sky. You can hear "robin's egg blue" and almost think only of that particular shade, though there's always a tiny flicker of eggs in a nest.

Smell hasn't been rendered so intensively metaphorical. We still have to say, "spicy, like nutmeg." We're still at the simile stage.

Why? Well, it's probably because we tend to talk less about smell. Illich maintains this is a culturally conditioned pattern, that this vocabulary was richer in the past and has become atrophied as our quest for a clean, deodorized, public space becomes more obsessive.

I'm not so sure. I think it may also have to do with the way our brains got organized in the ballooning of language and visual cortex. Language does expand when we pay attention to something. (It's not simply that we pay attention to something when we expand our language for it.) However, we're more inclined to pay attention to what we see—that's the default position for the brain. We snap to attention when we smell something different—smoke, stench, the aromatic pines in the ravine. But for the most part, air is scentless in a way that the world is never colourless.

"NO SIMILES, Doug. NO similes!"

The café windows were steamy on a cold autumn evening and the tables crowded. Canadian poet George Bowering wagged his finger at a member of the audience and delivered a brief lecture on flowery images before reading his own stripped-down lines.

Figurative language goes in and out of poetic fashion. Bowering was only making the same point that Shakespeare made four centuries ago in Sonnet 130—"My mistress' eyes are nothing like the sun"—where he poked fun at the unbelievable images poets used to describe their lovers.

"But *no* similes?" I thought, watching Bowering's wagging finger. "That would be like cutting out my tongue."

I HAVE FINALLY BECOME more tolerant of the need to make a distinction between metaphor and simile. They are not the mere residue of a grammatical construction that tacks on "like" or "as" without changing meaning. And regardless of what we

are often told in creative writing classes, not every simile
works better if you turn it into metaphor.

A simile is like the ≈ in mathematics: it says something isn't
quite equal to something else. Instead it approximates, it has
qualities in common. Saying "My aunt has grey hair like steel
wire" invites you see a certain quality of grim shine and stern
discomfort without imagining that someone's head is capable
of conducting electricity.

In fact, there are indications that the brain processes the
two figures differently.[21] For instance, one study looked carefully at how people interpret a metaphor like "Some ideas are
small diamonds" vs. the corresponding simile, "Some ideas are
like small diamonds." The former tends to be interpreted as:
"Some ideas are valuable and have a lot of potential for development." The simile, however, tends to be interpreted as:
"Some ideas are somewhat valuable, but they are small and
a bit disappointing compared with bigger diamonds." People
consistently make distinctions such as this between a metaphor and its simile form. It's as though each form invites
a different approach to creating a category in which the
compared items belong.

We also like to have an unusual comparison presented to us
first in simile form. We get novel comparisons more quickly in
this kind of construction, while the reverse is true for familiar
metaphors. In the first case, the simile structure signals that
we are to look for connections that we haven't made before.
It allows us to stand back and decide what features are to be
mapped from one side of the almost-equation to the other. We
don't need that scaffolding any longer when we encounter a
familiar comparison, and its presence only slows us down.[22]

RESEARCH regarding how brain function applies to metaphors and similes matters to me as a writer trying to write something that will reach out to other people. It confirms for me why I make certain decisions I've made in creating imagery. Working with metaphor is a process of finding the right balance between novel and familiar, general and specific. Often, the decision to frame a particular image as a metaphor or as a simile happens at the gut level, and sometimes it's based on something as simple as needing an extra syllable or two in a line. But looking back at the first few pages of my long poem, *The Office Tower Tales*, I see that yes, most of the simile constructions are related to more unfamiliar images—the planet going around the sun *like* a coffee cup off-centre in a microwave, silence dropping *like* cigarette ash. The reader needs the signal of simile to stop and look through the two layers of celluloid to see how they line up.

I love it when a novel comparison occurs to me—like when I walked out of an office tower on a day in early fall to see that one of the saplings planted across the street had suddenly changed colour, but only at the very tips of its green branches. It looked as though it had been given one of those eccentric hair-dye jobs that teenagers get. It was a young, hip, urban kind of tree, hanging around a downtown street corner.

But, as with robins and oysters, not every comparison a poet can make will work, and the range is relatively narrow if you want it to be accessible to significant numbers of people. I have never been a great fan of the paratactic approach to poetry—the jamming of different images up against each other, leaving the task of creating connections to the reader. This fashion grew out of avant-garde and surrealist tradi-tions in the early part of the twentieth century, and became one of the characteristics of L=A=N=G=U=A=G=E poetry in the late 1980s and early 1990s. Paratactic syntax pitches to that compulsion of the human brain to create some—any—meaning from language. Give us any fragmented jumble of words and we'll try to hook it into a sensible structure. Pickled cats' eyes and poetry? Okay, let's have a go...

Paratactic imagery does have its virtues. It strips out that frequently annoying narrative voice that says, "See? This is the moral of the poem. Here's why I selected these details" and leaves the reader some room to think, assess, breathe. But it's not a Holy Grail or magic formula. Parataxis can seem very trendy, but like most literary trends, it's not a new invention. It's simply a somewhat more extreme presentation of that metaphorical leap that we make always and continuously in language. The poet chooses two images and links them, either explicitly, with a little narrative around it, or starkly.

The paratactic gamble often doesn't pay off. Readers feel that some sort of head game is going on, that the poet is, in effect, saying, "I chose these fragments so that means I see the connections. If you can't figure them out, you're dumber than I am." Alternatively, readers may just decide that there is no connection and move on. Our brains quickly assess whether an external stimulus is meaningful or just noise; life's short, the incoming data is enormous.

I remember a positively certain young poet in a writing group saying, "I want to *challenge* my readers."

"Honey," I thought, "life's challenging enough. Give them a break."

So I go looking for images that will lead a reader or listener to say, "Yes, that's apt, that's right; that clicks for me, too, as a description of the world."

PERHAPS THE MOST profoundly important thing about metaphor is that it enables different narratives of the same events. An apt new metaphor can literally reconfigure the brain. New neural connections are made that will fire together for the rest of our lives. This, I expect, is the source of gut pleasure I get from a really good metaphorical linkage. I'm back in the early surprising world of a baby when every connection was new and our brains were in such a rush to make patterns stick.

Therapists know this rewiring capacity of metaphor when they try to pick up on the metaphors that patients use to describe their lives and offer alternative ones. Politicians know it, too, when they make up slogans like "War on Terror." This is not simply to make it "easy" to think about social responses to terrorism without thinking too deeply. Sloganeers don't want us to *stop* thinking about a subject—they want us to channel thinking in a specific way, recruiting the emotional linkages that are so fundamental to the process of metaphor.

Employing metaphor in slogans presents a great danger, of course. But it is offset by the fact that metaphor can change— that we can find all sorts of alternative patterns in the same material, depending on what we lay it over. For centuries, lyric poetry's metaphors for our planet were based on the images of an infinite resource: Keats and his "realms of gold" to be explored by stout Cortez; Donne's "Indias of spice and mine..." In our time, the planet becomes smaller, something to be tended and cared for, as in P.K. Page's poem "Planet Earth":

> *It has to be loved as if it were embroidered*
> *with flowers and birds and two joined hearts upon it.*[23]

From infinite expanse to embroidered handkerchief that needs tending is a profound restructuring of brains and it can be brought about by metaphors.

Perhaps the metaphors that we use to think about poetry and science could also be reconfigured. Roald Hoffmann, the Nobel Laureate in Chemistry who is also a poet, points out that our perception that the two activities are profoundly different stems from the metaphors we use to describe them. Poets "create"—a concept that involves putting pieces together. Scientists "discover"—a concept that suggests setting off across unknown seas (or looking behind the sofa). However, poets go shuffling through drawers or across the wide seas of literature to find ideas they haven't used before, and scientists build their ideas out of other components.[24]

Richard Feynman—scientist, not poet—once wrote, "What I cannot create, I do not understand."

"ALL RIGHT, we're going to put 'driving in the rush hour' together with 'knitting.'"

The seniors' writing group seemed bemused at first, and I wondered if it was going to work this time. It felt as though I was dragging them back to those horrible school exercises— "Think of a metaphor..."

I had primed the group by asking them to make lists of words associated with daily activities. When I asked them to connect two of the activities, knitting and driving, they wore those puzzled faces for a moment or two. But then they started to put things together—slowly at first, then with more speed and laughter. Potholes became dropped stitches. The red-green alternation of traffic lights became "knit one, purl one."

Trying to think of a metaphor is generally a sterile exercise in creativity. It works much better when you populate that comparison space in your brain with images and let combinations happen. And this happens at any age, though our facility does seem to change a little with age. I notice this when I do one of the wonderful exercises first developed by poet Kenneth Koch for school children.[25] It's beautifully simple. You write, "I have a coat of..." on the white/blackboard in front of the class. You say, "Now give me the weirdest thing you can think of to finish that sentence" and get the kids to bounce their ideas around. In grades three and four, you get lovely silly things: I have a coat of goats, of car engines, of rocks. By grade six, the responses have already tended to narrow, settle into categories. You get a lot of completions like: I have a coat of feathers... snow...fur. Things that logically can cover other things.

"What might wear a coat of rocks?" I ask the grade fours.

"A mountain," they say. "It would have bushes for buttons; it would have a snow collar."

"Brilliant," I tell them. "That's a metaphor. We've made up a metaphor."

All right, call it personification if you must, who really cares. What matters is reclaiming this birthright of language, of returning to the space where anything could stream into our infant brains to be associated with other sensations. The fundamental element of fun, of playing with the mouse under the rug and pouncing.

THREE The Holographic World

My new credit card arrived in today's mail.
They seem to get shinier all the time, and
I have now gone from *gold* to *platinum*
to *infinite*.

It has a glittery silver patch on the back, where a bird's raised wings shift colour through the spectrum. In strong light, you can see that the small image looks three-dimensional, the feathers overlapping as they fan out from the wing bone. The picture is a reflective hologram, a handy little piece of science that we use for fairly trivial purposes—CD covers, book covers, even special silver coins from the Royal Canadian Mint. Bigger and better holographic images have been created as art and hang in galleries—portraits that seem so genuinely three-dimensional that you are tempted to shake hands with the originals.

Clever gimmickry, illusion, a trick with the light. It seems interesting but essentially trivial, something to be used to make credit-card fraud a little harder or for a spooky, Halloween-like special effect.

However, some scientists think holograms may reflect much more than this—that the whole universe may be a trick with the light, an infinite credit card that buys us so much more than you'd think the limit would allow. It starts with the observation that you can cut a hologram in half but still see all of its image. I do not propose cutting up my credit card to see if this is, in fact, true. But isn't it an amazing idea, to take away half of the information and still have the whole picture? It will be fainter, not so well resolved, more ghostly. But both wings of the bird would still be there.

It makes you rethink the idea of completeness.

I CAME OF INTELLECTUAL AGE in a era that insisted the literary world was "partial," in both senses—that literature was biased and it was full of holes. The literary canon was tilted toward writers with a pre-approved credit rating. Great works of human creativity had been suppressed, more or less deliberately for reasons related to power. The record was full of holes and voices were irretrievably gone.

The holes were what became labelled the "Other," parts of human experience that we didn't want to face, acknowledge, understand. We were colonial, sexist and selective. Our brains were congenitally incapable of understanding experiences we had not had ourselves. There was no point trying to write about being a poor woman if you were a man, no point in writing about the experiences of an African slave if you were an affluent Westerner. You couldn't possibly get it right.

This world view seemed to be upheld by science—at least by a somewhat garbled version of the realization physicists had reached early in the twentieth century, that the act of observation always muddies the experimental results. "Aha," said the cultural relativists. "We told you so. There's nothing there except what we construct ourselves—and our constructions are inevitably flawed."

This way of thinking was foreign to someone like me. As a little girl, I had innocently enjoyed poems like Rudyard Kipling's "If":

If you can keep your head when all about you
are losing theirs and blaming it on you.
If you can trust yourself when all men doubt you
and make allowance for their doubting, too...

Reading it made me feel noble, as though I were standing with a falcon on my wrist, ready to toss it into the air. It never occurred to me that any of the language excluded me, even that last line: "And, what's more, you'll be a man, my son." It was simply a line to end the poem and make it rhyme and it meant I could aspire to being grown up, an adult.

But university studies thumped home that I was not supposed to like Rudyard Kipling or Hiawatha or even L.M. Montgomery. (I vividly remember the expression of utter scorn on my CanLit prof's face when I proposed doing a paper on her novels.) All these works had been revealed as partial and needed to be deconstructed. They could not be trusted. Large holes had opened in the paving.

THE IDEAS OF SCIENCE percolate as a thin stream into our awareness, through a coffee-pot liner of social context and trends. Science had a drip-drip-drip effect throughout the last century as thinkers tried to come to terms with the new para-doxes it expressed. Einstein's word "relativity"—unsettling enough—was overtaken by "uncertainty," a key concept in quantum mechanics.

Quantum mechanics had emerged in the 1930s as a way of understanding the world of the very small—the electrons and protons that made up matter, the photons that made up light and the wider spectrum of electromagnetism. It did not reveal a comforting universe. It was a world full of gaps. You couldn't measure certain pairs of characteristics, like the posi-tion and momentum of an electron, with absolute precision. This is not because we don't have refined enough instruments or because we muddy the purity of observation with our phys-ical presence, but because of something more fundamental. There must *always* be at least a little gap in how precisely you can know both these qualities at the same time. Usually this

gap is the tiny Planck constant, a mathematical sliver of uncertainty. However, the more you know about one quality, the
less you know about the other. If you pull out all the stops to
precisely measure an electron's momentum, your uncertainty
about exactly where it is located will increase without limit—
it could be anywhere in the universe. Measurement can only
provide partially correct pictures of the same event because the
dual properties can never be observed simultaneously in the
same experiment.

Nor could you ever know enough about a subatomic
particle to calculate which state it will transform into. There
is an essentially random quality to quantum events. There is
never enough information to let us to decide it will enter state
A instead of state B—only that there is a probability wave
associated with state A or state B. Faced with trying to relate
the undeniable effectiveness of their equations to a "real"
world, physicists ran into a bog. What is probability a wave *of*?

"Who knows?" said scientists like Niels Bohr. "Don't ask. We
can tell you the probability that a certain event will happen,
and, to do so, we use the same mathematics that we use for
describing 'real' waves of light or water. But we cannot tell you
what's waving."

The quantum math shimmered like the foil dove on
my credit card. Probability waves buy us everything from
laser technology to the modern cryptography that protects
Internet banking. But, said Niels Bohr, mathematical language
could not be translated into other human languages in any
meaningful way.

When this "Copenhagen interpretation" of quantum
uncertainty emerged into public awareness in mid-century, it
prefigured the postmodern push for alternative narratives. All

languages are like mathematics, theorists of linguistics and culture said. Each is a self-contained system—it doesn't have to refer to a real world to be effective. And there are holes in what any language *can* say.

I first read about the uncertainty principle during the 1970s, in *The Tao of Physics* by Fritjof Capra, one of the earlier books to popularize the ideas of twentieth-century physics. The title reflected its times—an era of young people (and not-so-young rock stars) exploring ashrams and Rolfing, when down-home Western society was coming to terms with the fact that the planet's cultural geography was both wider and closer than we'd realized. In that time of Indian muslin skirts and Nehru collars, *The Tao of Physics* imbued the paradoxical results of the double-slit experiment with the air of a Zen koan.[1] Uncertainty and its inherent unknowability took on a slightly warmer cast than it had under the Nordic light of the Copenhagen interpretation. In my philosophy class at university, the random nature of quantum decay became a possible source for free will and human freedom to choose. And science, argued Capra, was becoming one with the great oneness underlying Eastern religions:

> *Physicists have come to see that all their theories of natural phenomena, including the laws they describe, are creations of the human mind; properties of our conceptual map of reality rather than reality itself...All natural phenomena are interconnected, and in order to understand any one of them we need to understand all the others, which is obviously impossible.*[2]

THE REASON that a hologram can be cut in half but still give you the complete image that was photographed in the first place is that the information encoded by light doesn't go from subject to film in a simple point-A-to-point-B line. It is transferred by treating the light as waves rather than particles. The film registers the interference fringes where light waves interact, not single points where a photon arrives. As a result, information is broadly distributed on the film's surface.

The dove on my credit card is produced by a film soaking up light waves on its flat surface, where the information from the photographed object has been projected from three dimensions down to two. If you could see the film itself where this information was recorded, it wouldn't make much sense. The data is smudged and smeared the way waves are when they muddle around the pilings of a dock. But when the information is printed on a thin reflective sheet of aluminum and light shines on it, your eyes can reconstruct the roundness of the original object from the data about the wavelengths.

Three-dimensional holograms also exist: images you can walk around to see front, back and sideways. Here, the interference of the light waves has been mapped to patterns in the higher number of dimensions.

Any point on a hologram can carry information about more than one thing. You can even print holograms over top of each other on the same film. The image you reconstruct later depends on the light you shine on it—it has to be the same kind of laser light that produced the image in the first place. But all the pictures are there.

HOW CAN A PART know a whole?

Central to answering this question is the concept of "information": What is it? How is it transmitted and transformed?

Today, there are shadows on the wall beside my desk, cast by the sun on an equinox afternoon. Tendrils of golden clematis cast scrolled lines on the floor, like delicate cantilevers reaching for the other side of a sunlit gap. Somehow you notice shadows so much more during spring and fall, when light's angles change. But, lovely as the shadow's lines are, they are much less than the brilliant originals draped outside my window. The shadow-flowers are mere blotches compared with the brilliant pointed-pagoda petals of the real flowers. So much detail is lost when information is projected to my two-dimensional floor.

This is such an ordinary observation that we assume it is an underlying feature of reality. Plato's metaphor of the Cave is central to much of Western thought—we experience a "real" world as if we were seated in a cave with a fire behind us and could only watch its shadows on the wall. The world is immeasurably poorer than the "original" one that generates its shapes.

Contemporary physics may turn this idea on its head. The idea of physicists like Gerard 't Hooft and Jakob Bekenstein is that the universe itself may be the projection of information on a surface (a boundary) into a higher-dimensional space.[3] We have a world of three dimensions and time, fundamental field forces like gravity and particles like electrons and photons that move around, all because somewhere in the background is a lower-dimensional framework where everything is reduced to its most minimal essence.

The boundary is made up of units so tiny that they are neither space nor time nor force nor matter nor anything else.

They consist of a single piece of information, a single "degree of freedom." (To a mathematician, a degree of freedom is any quantity that can vary, such as a co-ordinate along an axis in space or the on/off state of a transistor in a computer chip.)

A holographic universe reverses Plato's idea that information is made poorer when it is projected. The limited degrees of freedom on the boundary can be recombined into much richer degrees of freedom in our experienced world.

AT ONE LEVEL, language has very limited degrees of freedom. It's made up of phonemes, sounds, basic physical things, linked to very precise positions of teeth, tongue, vocal cords. Certain combinations are routinely permissible. And yet they are essentially abstract bits of information. What is the sound "d," for example? It's something that is common to "dam," "endogenous," "dentist" and "drywall," but it has little inherent significance. You cannot, in fact, even make the sound by itself. When you take a spectrogram of a sound like "duh" and cut away the portion that corresponds to the vowel, you eventually get to something that sounds like a glide or a whistle—not a human speech sound at all.[4] This is because at every instant we are transmitting parallel information about both the phonemes, "d" and "u." The sound of "d" has only the tiniest possible degree of freedom by itself.

However, when you project the bits of sonic information into more complex dimensions—words, sentences, poems, literature—their degrees of freedom expand dramatically. The same sound can carry many different kinds of information. The "d" in "and" serves a different function than the "d" in "death" or "blood"; it is a patch on the hologram that can be illuminated by different wavelengths. Take for example, this stanza from Gwendolyn MacEwen's "Magic Animals":

But the mandrill with his mardi-gras mask
folds his arms and examines a world
more surreal than his rose-red ass.[5]

Sound's slow waves glance off the reversal of "d" and "r" in the first line, glimmer in the close association of "d" and "l" in the first and last words of the second line, then focus in on the hard, almost stand-alone "d" in "red," just before that final semi-rhyming "ass." It is both a point particle of sound and part of a shimmering net where different point particles are linked over and over in multiplying combinations.

THE IDEA of a holographic world addresses one of the oldest questions in philosophy—is the universe one thing or many? Can matter be subdivided down to discrete atoms, a smallest thing and then no further? Or is everything somehow part of one thing—in other words: is there a seamlessness, a continuity underlying all those discrete particles?

Physicists pondered this old question quite a lot over the last century. At first it seemed that they were on their way to finding out that there were smallest things—the atom divided into quarks and electrons, light was chopped into photons. In the words of Rodgers and Hammerstein, we'd "gone about as fur" as we could go.

However, throughout this chopping process, there was an ongoing problem with waves. They don't act the same way that point particles do—they have a continuous quality. And it became apparent that everything from electrons to elephants

has a wave-like character, though we can't see it manifested
in something as large as an elephant. However, this continuity
became a mathematical headache. If equations assume that
space is continuous and infinitely divisible, they eventually
explode into singularities, points where results go off the
charts into unmanageably infinite quantities, a devastating
quantum foam.[6] So physicists have gone back to developing
approaches like string theory and loop quantum gravity to take
the smallest things idea down a step further. These theories
posit that the fabric of space-time is made of tiny vibrating
strings or loops. Most recently, the smallest things idea has
been worked further, bringing us the picture of a holographic
universe made from surfaces where infinitesimally small
degrees of freedom project back into a continuous three-
dimensional world.

A holographic universe sounds strange, but that's not surprising in any theory that attempts to account for a universe
where some very strange things have actually been observed
in the last thirty years of experimentation. The smallest things
we've been able to isolate—such as photons of light—aren't,
in fact, isolatable. Two photons that have interacted with each
other remain entangled, regardless of how far apart they fly.
Measure a characteristic of photon A, and you can affect photon
B, even though they no longer seem to have any way of communicating.[7] Send a stream of photons through a filter in the
shape of the letter "A," send their entangled partners off in
another direction entirely, and you'll record a ghostly letter "A"
when you catch those partners on a screen, even if there is no
way the information could have been transferred.

The universe communicates with itself in ways we do not
yet understand.

WRITING OFTEN HAS a holographic feel to it. In recent years, I've bought Lucy Maud Montgomery's journals as volume after volume was published, reading them with the same avid intensity that I read *Anne of Green Gables* when I was a little girl. Her journals recreate a whole world (the microcosm of a church parish, the macrocosm of world war) with its vivid backdrop of social and technological change. Newly invented motor cars purr along the back roads of Ontario, literary fashions change and leave her popular success in their wake, new (now quaint) treatments become available for treating mental illness and physical pain.

The journals can be almost oppressive to read, with the incessant work, work, work of a driven woman: Missionary Band and Sunday School Concerts, pickling hams, putting up pears, and putting up with visits to parishioners. The idea of a life where you might get half an hour to yourself at the end of the day makes me wince with a combination of guilt and pity. The pace she maintained would drive me literally insane in a month—especially when combined with the emotional stress she was under. In one horrible year, she lost her beloved cousin Frede to the Spanish flu epidemic, found that her husband had developed "religious melancholia" and wrestled over ongoing copyright issues related to *Anne of Green Gables*—all of this with no apparent emotional outlet, no capacity to discharge the stress safely except within the pages of her journal.

It seems curious, to a woman today, that the prospect of Ewan losing his parish would strike such horror into her. The idea of having to put him in a sanatorium, break up the Leaskdale manse and find another home was unthinkable, and she exerted all her formidable energy to conceal and make up for Ewan's illness. But why? She was a well-known writer who could make a substantial sum from her pen. Leaving the duties and constraints of a minister's wife's life should have seemed like a back door into freedom. But she didn't seem ever to put her hand on the doorknob, even in thought.

Perhaps the financial constraints of raising two boys on a writer's royalties would have been too much. However, the real barrier to freedom was psychological. Her upbringing, the expectations drilled into her as a girl and reinforced by a young womanhood spent in cramped attendance on her grandmother, engrained a sense of Protestant duty combined with a deep-down sense that a woman derives her status and place from her husband. No matter how successful she might be in her own right, L.M. Montgomery could not really exist outside the confines of a book jacket. Her real social identity was that of Mrs. Ewan MacDonald. In that identity, she was mother of her sons; in that identity she decorated her home and, in fact, had a home to decorate.

Her journals make you feel you are getting everything about her life. We might want to say that this is illusion, that the construct we are being offered is only a partial one, that there are many other narratives that could be made about her. But this is unsatisfactory to anyone who reads the journals. The act of reading them is like fingering a branch of coral, the solid carbonate structure that remains when the myriad minutiae of its active life have gone. It may not be complete, but what remains is real. And it is more complete than even she knew. Maud Montgomery had sometimes lamentable attitudes about class and the French people of the Maritimes; she was misguided about child-rearing and mental illness. She did not think to hide certain attitudes and assumptions that seem strange or even unpleasant today, because to her they were invisible. You can't hide what you can't see.

As Walt Whitman wrote, "If you love to have a servant stand behind your chair at dinner, it will appear in your writing—or if you possess a vile opinion of women or if you grudge anything...these will appear by what you leave unsaid more than by what you say."[8]

The ghost always appears on a screen somewhere.

OF COURSE, a memoir is intended as a more-or-less consciously complete picture of the world—or at least of the bits of it that the writer is attending to. However, the sense of completeness generated by a work of literature applies to all its forms. For centuries it has been one of the main goals of writing, and it remains the source of much of its pleasure. Think of the world created in the Anglo-Saxon poem, *Beowulf*, with its Shield Danes and monsters, a heroic society made real by its accoutrements and funerals. Or Dante's complex fantasy of hell with its characters and detailed geography.

Twentieth-century theorists maintained that creating this sense of seamlessness in art was, in fact, playing into a dangerous human tendency toward totality. They actively rejected the idea that the poet could or should present a comfortingly coherent view of life. Novels and poems of collage and fragment were better for us because they constantly remind us that nothing *is* complete.

However, the pleasure we take in the completeness of a work of literature is not just laziness, naïveté or a mindless submission to authority. It's a very active, physical pleasure, probably related to the satisfaction we take in constructing a coherent picture of the world around us. Our own consciousness arises from the mental sub-circuits that take in information and create an apparently seamless whole. Cognitive science has done a great deal in recent decades to probe the way in which the brain integrates perception—not through a little rational homunculus, who watches a theatre of the mind and makes sense of it all, but through the constant shifting of circuits for attention.

We know there are many holes in the mesh of perception—that, for instance, the eye isn't really seeing what's in the blank space on our retina where the fibres of the optic nerve channel themselves away. We fill it in and are never aware of the gap, as we are unaware of many other gaps in the branch of coral.

But the idea that "illusion" equates with "unreal" is an unnecessary leap to make. There is no particular reason to think that this sheet of paper that I write on isn't continuous because at any point in time there is a spot on it that is invisible to me. To insist that, because you can't say everything, what you *can* say is erroneous or lacks coherence—well, that's not how a holographic universe would work.

"ALL NATURAL PHENOMENA are interconnected, and in order to understand any one of them we need to understand all the others, *which is obviously impossible*," wrote Fritjof Capra (my emphasis).[9] But physicists do not in fact assume that you must include everything in order to make meaningful descriptions of the world. There are many ways around the need to describe *every* degree of freedom *all* the time.

For instance, there's renormalizing—a way in which mathematicians can reset a quantity in their equations to zero when it would otherwise lead to unworkable infinities. When renormalization was first developed as a mathematical technique, it was viewed with suspicion as a trick, a cop-out, a "shell game" in the words of Richard Feynman.[10] Could you be sure your results were accurate if you left out the troubling infinite quantities?

But physicists have generally become more comfortable with the process. "Renormalization is not such a bad word," writes Lisa Randall. "It refers to the fact that at each distance scale of interest, you pause to get your bearings. You determine which particles and which interactions are relevant at the particular energies that interest you at the moment."[11] She compares it to the way our eyes automatically average the shade density for small areas when we are viewing something with fuzzy resolution—in effect, we choose a pixel size that works for what we're trying to look at.

Consciousness is constantly choosing the appropriate pixel size. It doesn't usually pay attention to the mechanisms of oxygen transfer in the blood, unless we decide to look at that process in a blood test or an electron-scanning microscope. Consciousness makes the assumption that small things are built coherently and consistently into bigger things; and the fact that the world's smallest components are entangled gives this assumption some street cred.

The concept of a holographic universe is another way of making data manageable. It is, in spite of the complex mathematics needed to describe it, a way of economizing on information. As physicist Lee Smolin puts it, "If the world really were continuous, then every volume of space would contain an infinite amount of information. In a continuous world, it takes an infinite amount of information to specify the position of even one electron."[12] By contrast, the idea of a holographic universe puts a limit on the amount of information required to describe three-dimensional space—and that limit is not related to the *volume* of space but, paradoxically, to a kind of *boundary* that encloses it. These boundaries (or screens) can be thought of as a kind of channel through which information flows from one region of space to another, with the area of the screen's surface as a measure of its capacity to transmit information. The screens make up a network of holograms, each containing coded information about the relationships between the others.[13]

It is as though shadows cast through two different windows combined to give me the brilliant gold tangibility of clematis flowers.

THE HYPOTHESIS of a holographic universe involves a nicely paradoxical reversal. Instead of underlying depths providing the basis for the world of surfaces we observe, it is the surface that underlies the world of extended dimensions where the fish can swim or doves can fly.

On my credit card, the hologram dove floats a few micrometres thick; it is superficial, decorative. Our proverbs instil in us a deep distrust of such appearances that are only skin deep. "All that glitters is not gold," we solemnly warn ourselves and remind ourselves to value what is deep and gets to the heart of things.

Perhaps we need the warning because human beings seem naturally drawn to glitter. Two hundred thousand years ago our ancestors were carrying pretty bits of rock hundreds of

kilometres away from their origins, for no apparent reason
beyond their attractive surface. Ellen Dissanayake points
out the widespread preference for "the quality of brilliance
or dazzle" in cultures around the world. Sheen, gloss,
scintillation—we are all drawn to the quality that the Yolngu
people of Australia call *bir'yun* and create by adding hundreds
of tiny dots of white or yellow to their paintings.[14] *Bir'yun* is
not simply the shimmer of health and well-being, but also
extends into associations with joy and sacredness—a bit
different from the idea that sacred things ought to be deep.

Deep thoughts are often rather simple, even banal,
especially by the time they've been noticed by a large number
of people. That space and time are joined was a particularly
insightful understanding reached by Einstein; that human
beings are part of a continuum of life was a deep contribution
by Darwin. But such thoughts don't make poetry (or even
scientific theories) without a great deal of surface detail. The
deeper ideas that poets may want to present—ideas about
breath and consciousness, about justice and freedom, the
nearness of death and the need to look after the planet—
aren't all that different from one poem to another. It's a "nice"
surface that distinguishes poems, not their ideas.

INTO THE WORLD of partial literature comes Miriam. She was
part of my writing group for a time—a bright, passionate girl
from the streets of Santiago whose wings had blown her to the
northern plains of Alberta. She had never had much of a chance
at education, but she was desperate to express herself, even in
the cumbersome cold of English.

We kept recommending books to her—novels by writers like Isabel Allende or Gabriel García Márquez. We assumed that writers from her own part of the world, writing in her own language, would have the most to say to her. One afternoon, Miriam phoned me in a passion that was just this side of tears, to tell me about the wonderful book she had just finished.

"He has written *my* story," she kept saying, caught up in a tide of something that was release and close to resentment.

It was not by Allende or Márquez. Instead, it was by a white male of a completely different generation, set in a completely different part of the world—*Angela's Ashes*, by Frank McCourt. It reached across the apparent chasm and held her by the throat.

As for Isabel Allende? "What does she know about life on the street?" asked Miriam scornfully. "I know what it's like to be poor. So does he."

This is the completeness of good writing, that its partiality can be trusted.

FOUR Points on the Line

I loved drawing straight lines when I was a kid. My ruler, a foot-long length of sturdy hardwood with a thin strip of metal inset along its top edge, described an unwavering commitment from one point to a second one. Even on days when I couldn't find the ruler and had to use the nocked edge of my protractor (chewed by the dog) or the pebbled cover of my math book, the satisfaction of that ideal line shone through its bumpy manifestation on real paper.

You feel you know where you are with a line, which may be why the image of the line has been so prominent in Western thought since Aristotle formulated his "golden mean" stretching between two extremes.[1] Aristotle taught that virtue meant finding the appropriate balancing point on the line between, for example, recklessness and timidity. Ever since, we have conceived of a line as the distance connecting distinct polarities. Nature/nurture. Body/spirit. Wave/particle. Male/female. Political left/right. Good/evil. Our ideologies assume that there are two things, two *different* things, however much we find that one shades imperceptibly into the other.

Dualities create controversy. They are the things we argue most intensely about. ("Are people naturally bad or are they made that way by their environment?") Even poets can be surprisingly argumentative about the dualities we work with: sound vs. meaning, prose vs. poetry, superficial vs. deep. We are hardly ethereal in these debates; we become heated and occasionally nasty. But that's because they are not just debates over the relative importance of free verse versus metrical forms, but over virtue, over what we are doing with our lives.

TWO YOUNG MEN are standing before a single microphone and making strange noises. The first of them utters an extended buzz—zzzzzzzzz—while the other launches into a series of clicks and gargles. Back and forth goes their bizarre duet. Many of the audience in this below-ground bar are laughing aficionados; others look completely bemused or even indignant. I know how they feel. I remember the first time I saw this happen at a poetry reading and wondered what the hell was going on. "It's all right," I want to tell them. "Relax. Just go with it."

Jeff Carpenter and Glenn Robson (aka Tonguebath) are sound
poets, inheritors of a tradition that goes back to the 1970s
but has links with much older traditions from chant to scat
singing. Sound poetry tries to focus entirely on the sounds of
language and not language's meaning.

In a sense they're doing what I used to do as a kid when I
would stand on a small patch of daisy-freckled grass beside
Nana's house and spin slowly with my arms out, saying a word
over and over again. *Work...work...work...work...work...* At a
certain point, the syllable would disaggregate and become
a series of sounds. Nothing tied them together into a word;
nothing looped them to meaning any longer. It was a weird,
disorienting and mildly addictive place to find in my brain. And
it's surprising how many writers I know who say they did the
same thing. Sound poets just do it in public. As grown-ups.

It seems to be a guy thing—I know very few seriously
dedicated female sound poets. This could, of course, be cultural
conditioning. Maybe I no longer stand in the garden with my
arms out saying "bird...bird...bird" because women are under
stronger societal pressures not to look silly to the squirrels. But
I personally think it's something on the Y-chromosome. Even
non-poet males seem to like making silly noises. While, if the
impulse has been crushed by cultural conditioning, the job has
been so thorough that I can't even find its crushed remnants
in my psyche. I'm certainly not actively suppressing an urge to
hoot.

EVEN A SOUND POETRY PERFORMANCE can't eliminate meaning
entirely. We never experience sound without assessing whether
it has meaning, regardless of whether the noise is language or
not. Our brains evolved to scan the environment constantly for

significance. Is that click in the night something to worry about? Some natural noises come at us with no intended meaning—wind roar or water gurgle—and we rapidly assess them as non-communication. (Although we still attribute a kind of emotion to them—our own emotion at what the sound implies for us. So the wind's roar "means" discomfort or the appreciation of shelter, and the water's gurgle "means" peace, tranquility.)

But it is very hard for us *not* to search for intended meaning in any animal noise. Even the wordless drumming of fingernails on a table communicates. It takes a lot of energy for animals to make noises. They don't do it casually.

So the buzzes and growls as the Tonguebath duo played with consonants and vowels couldn't come without some sense of intentional significance. Of course, a lot of that was non-verbal—the antics of their faces and hands that said, "It's okay, you can laugh." But still, the noises themselves can carry a freight of emotion. That *zzz-sssss* poem had all the resonances of snake. So even a "pure" sound poem doesn't get away entirely from meaning, which is partly the poets' point.

Our brains process sound and meaning in a simultaneous braid of sonic analysis.[2] Input from the nerve cells of the cochlea is immediately fired off to a series of structures in the brain stem, where it is mapped and analyzed for location and frequency. A series of filters sharpen up what we're hearing; some neurons fire specifically when sounds start or stop.

Meanwhile, neuronal impulses are travelling a second set of pathways, the diffuse ascending system. This process is not as understood as the previous one, but it communicates with the parts of the cerebral cortex responsible for attention, memory and learning. The diffuse ascending system seems less concerned with identifying sounds precisely; the neurons respond a little more slowly, as if they were averaging sounds over a period instead of reporting the latest event.

As sound streams in—be it screech, sonnet or symphony—the brain is constantly anticipating what might come next, confirming or modifying its guesses about a sound's significance and relation to other sounds. With language, the tiniest components of sound carry meaning—the "s" that makes a word plural or the "ing" that prolongs an action. Or the fact that certain sound combinations signal the beginning and end of words—think of the "md" at the end of "slammed," which could never occur in English at the beginning of a word. These morphemes all create a web of potential significance that we lay over the actual noises coming in.

So sound and meaning are not duelling dualities at either end of a line, battling to exert the greatest weight while poets try to find the virtuous point where they cancel each other out. Nor are they the ends of a teeter-totter—when one's up, the other's down. A line of poetry is not a thin jointed string of words but a beam that transfers force because the qualities of sound and meaning are entangled along its entire length.

"THE SHORTEST DISTANCE between two points is a straight line." This familiar wording encoded Euclid's first axiom for generations. My father often told the story of the day he learned it indelibly, in a Depression-era school in the rundown Vale of Leven. The teacher chalked the theorem on the board, turned to the raggle-taggle class nourished on jam sandwiches and stolen turnips and said bitterly, "I don't know why I'm

teaching you this. You're only going to grow up and dig ditches."
As though all their lives, including his own, were hopelessly
straight lines to unavoidable ends.

But even then, back in the 1930s, mathematicians had
been redefining the idea of a straight line for nearly a century,
after Georg Bernhard Riemann developed the concept of non-
Euclidean geometries. On the surface of a sphere, the shortest
distance between two points is *not* straight as my metal-edged
ruler would have drawn it. It's a curve, the arc of a great circle.

Still, even in alternative geometries there is something to
the idea of straightness. There is a "shortest" path, the direct
route, a way of getting from A to B that takes as little energy as
possible. I like the idea developed by physicist Richard Feynman
that such a path is a "sum-over-histories."[3] In other words, the
path travelled by a photon or an electron between two points
in space-time includes all possible paths anywhere in the
universe. But for every path that goes up, another goes down,
for every one that zigs to the right there's another that zags to
the left. So, mathematically, the various routes cancel each
other out and all these virtual paths add up to our familiar
straight line. Nevertheless, the ghost of those alternate paths
is a kind of aura around the line, making it a rope, a cable,
rather than an infinitesimal one-dimensional abstraction.

This reminds me of a remark T.S. Eliot made in one of his
essays. He was talking about the sonorous vagueness of certain
Victorian poets like William Morris who, however suggestive
they seem, "really suggest nothing; and we are inclined to infer
that the suggestiveness is the aura around a bright clear centre,
that you cannot have the aura alone."[4]

He is saying, of course, that the meaning of a line of poetry
is more than a straight line—it's a sum-over-potential-

meanings through all the echoing pathways words might take in our brains. But a poem is not allusive for the sake of allusiveness. There is a path at the centre of it that is expected, a straightest line.

THE ASSUMPTION of meaning's straight line got decidedly blurred by the middle of the twentieth century. Another dichotomy emerged to get people excited: between language as conveyor of thought and language *as* thought.

The older picture had been, roughly, "I've got a thought in my brain that I want to communicate. So I'll fish around to find the right words for presenting it. You'll hear the words and the same picture will be triggered in *your* brain." It was the basis on which human beings had manipulated and negotiated language for thousands of years. Behind it lies another assumption, that there is an objective world out there to have ideas about.

But suddenly this straightforward line from speaker to listener became naive, out of date. Language doesn't translate thinking—it *is* thinking, said theorists like Jacques Lacan.[5] There isn't a pre-existing set of concepts in our brains that we match up with words. And the language you happened to learn as an infant conditions your thought processes so thoroughly that it determines what you *can* think. Speech is not a transparent medium through which we present truth about our internal thoughts. Instead, it is a stained and twisted construction coloured by unexamined assumptions about the world and the society in which we find ourselves.

The extreme postmodernist position is that language creates thought rather than the other way round. There isn't anything going on inside our brains that *isn't* language—it is the central experience of our lives. Meaning must be constantly decon-

structed so we can examine the implicit assumptions underlying a word like "actress" or a phrase like "war on terror."

Theorists based such conclusions on evidence from anthropologists and linguists that showed the startling differences in
how languages around the globe are structured. The apparently
obvious subject-predicate relationships of English and its
European cousins don't at all reflect how Native languages of
North America or the tongues of the Far East are structured.
Verb tenses and "I-you-they" pronouns so central to English
are anything but universal. The arbitrary, culturally determined
aspect of language was summed up in factoids: Inuit languages
have fifty words for snow; Gaelic doesn't even have a word for
"no."[6] Early in the twentieth century, the Sapir-Whorf hypothesis even put forward the startling idea that the language
you speak conditions your experience of space and time.

So, if there is no common pot of thoughts in our minds,
argued the postmodern thinkers, then communication isn't
about making the same thoughts resonate in someone else's
mind. The path from your brain to someone else's is not
straight and necessary; it is a random accidental leap. What
you take out of a text doesn't reflect the writer's intentions but
your own. Language had become *texte*, a constantly rubbed-out
and written-over blackboard that we spend our whole lives
writing and rewriting, like students left alone in a classroom
after school with no teacher. And in the late 1960s, Roland
Barthes went so far as to claim the death of the author.

This assertion, of course, makes many writers like me cross
and quite argumentative. I feel like the character in the Monty
Python sketch about the Black Death, being hauled out to the
cry, "Bring out your dead."

"I'm not dead yet," he moans. "I want to go for a *walk*."

What do you mean I don't know what I'm saying or what I want to get across to a reader? What do you mean I can't do that? This statement seems to undermine my whole life.

THERE HAS BEEN an analogous debate in mathematics over the past century over what physicist Eugene Wigner termed "the unreasonable effectiveness of mathematics" in describing the world.[7] Is nature inherently mathematical, or does it just seem that way because we use mathematical tools to probe it? In other words, are its regularities an artifact of the language we use? Lurking in the background of this debate is the same assumption called into question by the linguistic determinists: is there an objective world out there for us to have ideas about?

To worry about this question at all means you are subscribing to one of the oldest dichotomies in Western thought, between "mind" and "body." This duality assumes there is something inherently and essentially separate about mind as it witnesses the world. But you can only get fussed over the possibility that language (mathematical or poetical) is an imposition of mind on the world if you believe they are genuinely different. If you assume your brain is part of the world, that it is made *of* the world, then the languages it creates respond to regularities that are features of the same combined system.

I REMEMBER reading a letter from Jacques Derrida to a Japanese colleague on how to translate his influential term "deconstruction."[8] Deconstruction, he seemed to be saying, is something that does not take place in time—it's not an event, or a process, it's somehow something that happens for all time. It doesn't happen in space either; it's a breaking down of the ordinary structures between centre and margin, up and down. And it's not something that can even take place in language; at great length, words fail him. I had the sense of someone from a Borges story chasing a shadow down endlessly receding corridors.

This will-o-the-wisp is not about writing in any sense that I undertake the act. Language is a biological activity. That is its limit and its glory. It cannot escape time and space, and it certainly can't escape having a mind at either end of the process. It is true every reader and every age will experience a given work differently—true, but only trivially so. As human beings, we can't help but understand a piece of communication as coming *from* someone *to* someone. It therefore takes place in time and in space; it is an action, with consequences. It's not going to float off into some semiotic space, spaceless, timeless and inconsequential. We're animals; we're not going to go to all that trouble to make noise unless it's worth the bother, unless it's meaningful, unless it has results in the world.

I DON'T THINK it's recognized that the postmodern view of language is largely an artifact of a particular view of the human brain—the behaviourist view of the brain as a black box. Most postmodern theorists would be horrified at being linked in any way to behaviourism, which they viewed as a dreadful leftover of the scientific, deterministic, single-narrative modernism they were reacting against. However, in an unexamined way, a behaviourist picture of the brain was the basis for much of their argument.

To the founders of behaviourism like B.F. Skinner, all you could really know of the brain is what goes in and what comes out. Certain stimuli result in certain outcomes, but the brain

in between is invisible and inaccessible. It was visualized as
a kind of aspic that can be shaped by any cultural stimulus it
encounters, a "blank slate." It follows there is no necessary
commonality between how any two brains process a stimulus. You see a certain wavelength of light go in, and the word
"red" comes out in response. But what happens in the mysterious middle may be entirely idiosyncratic, and my experience
of red may be entirely different from yours. This is the question philosophers refer to as "qualia" and debate with unabated
vigour throughout the ages.

It's not hard to see how this picture of the brain can also
underpin the postmodern view of thinkers like Derrida, who
imagine literature as a kind of transmission grid, in which
"texts" get pumped in from various generators and the energy
extracted by different users, without any specific one-to-one
flow of electrons from point A to point B. To postmodernists,
language is the outward and visible sign of invisible, individual
and idiosyncratic processes. There is no commonly agreed-on
narrative that they necessarily embrace, any more than there is
a common process for producing those particular sentences.

I sometimes think the postmodern position was the
product of highly articulate, literate people so accustomed
to manipulating language that they didn't have any inkling
of thinking or consciousness without it. And, of course, in
human society (particularly Western society), people with high
abilities in the manipulation of language tend to be the most
successful. So, ironically, the postmodern foregrounding of
language may be a function of hierarchical Western structures—
the very thing that postmodernism supposedly subverts.

True, even those of us who are not semiotic theorists find it
hard to get away from the constant natter-natter-natter of the

language loops in our busy brains. But now and again I can just about get there. Watching a baby stare at a revolving mobile with intent eyes: a world of observation that operates without words.

Or, a gardener looking at green.

There is, in fact, a whole set of activities in our brains that is not language. There is a state of observation and a state of talking about observations—there is a start point and an end point and a path between, where observation evolves into thought and only then into language.

OF COURSE, my assertions about language and meaning have no more inherent legitimacy than John Watson's or Jacques Derrida's. What justifies a change in thinking is being able to look inside the black box at last. Even with still-crude tools, we can see that brains are not blank blancmange but highly structured systems with a good deal of commonality in what they do and how they do it.

We can see the language loops light up when the brain is presented with words; we can see the processing of written language and how it differs from processing oral language.[9] We can see the intense activity going on in other parts of the brain—not just in the language section, which sometimes does shut up and take a silent back seat.

And the science of linguistics gives us a more nuanced understanding of the coarser findings of a century ago. Yes, languages differ in subtle and fascinating ways. Take, for example, the speech of the Tzeltal people of Mexico, who don't use terms analogous to "left" and "right" to orient their world but instead describe spatial arrangements relative to the mountain slope that dominates their villages. Even an arrangement of objects on a table is "up-the-slope" or "down-the-slope."[10] However, the differences are an aura around a common thread of bodily experience that bounds all languages, ties them to a world of up-and-down gravity, unfolding time, solids and liquids, seasons and emotions. It's not quite true that the Inuit have fifty words for "snow"—it depends on how you define "word," and English can discuss all the variations of snow, just as an Inuit poet could write about an elephant if she needed or wanted to.

STILL, every poet has faced the fact that the line of meaning can be anything but straight. You really don't know what people will make of something you've written.

A while back, I found a discussion of one of my own poems online. It began with the plaintive question, "Does anyone know what this poem means?" The poem, called "Thief," had been picked up from my first poetry collection for a school anthology. It dealt with the aftermath of having my house broken into, which is supposed to leave you with the sense of violation when your belongings are pawed by strangers. In fact, I had felt a curious sense of relief. I realized that what thieves want to take isn't what I most treasure. But, of course, what I most love *will* be taken from me, and I can't keep the great thief Death out of my house with padlocks and burglar alarms.

This was what the poem "meant," so I was amused to read:

What i think so far is that the writer is talking about
someone that she's in love with and that person cheated on
her before, and now she's afraid that her lover will cheat

again, yet there's nothing she can do about it because she's
in love with him. "No care that I can take, no shutter on my
heart, can keep you out"

By the line "Your hands hasty, random, stupid for what
they take and what they leave behind"...i think it means
that she thinks she's better than the other woman her lover
cheated on her with.

And then a further despairing cry: "If you find any metaphors, personification or anything else PLEASE let me know. Btw...i got all the alliteration and stuff..."

No, said another respondent, that's not what it means. She probably suffers from heart disease and has gone to the door of death but survived.

The interpretations made me laugh—my cardiac condition, physical or emotional, was not at issue in writing the poem. At the same time, they touched me and reminded me how meaning gets tangled up with assumed biographies. The one personification in this fairly simple poem is Death, unnamed but characterized by "your scythe-shadow on the wall." However, even that stereotypical detail slipped past the notice of the first reader, and the blanks got filled in with a personal narrative that may well have had more to do with the reader's life at the time than mine.

This all supports the notion that readers construct their own reality. However, the elephant in the postmodernist parlour has always been that, to a surprising degree, we do understand each other, even across cultures and languages. Take that plaintive query—"what does this poem mean?" In fact, most of the posts on the discussion board did "get" the poem. The back-and-forth chitchat converged on a description of what the poem meant that doesn't differ much from what I intended.

As humans, we feel a great deal of satisfaction when we feel we have decoded the meaning of a string of words and feel we "understand." Postmodernism had seemed to threaten that basic satisfaction, especially when poets took up its challenge with enthusiasm and created several generations' worth of poems that withdrew as much as they could from the personal narrative of the poet's eye, leaving words lying around like Lego blocks to be assembled into what the reader could make of them. Readers persistently declined to be grateful for this freedom to build their own Lego constructions and kept looking for the writer's architectural blueprints.

THE WEB CHATTER about my poem is absolutely typical of how readers engage with a narrative. In his book on how readers actually read, David Miall cites research demonstrating that readers feels engaged with the narrator of a poem or story as if they are partners in a conversation.[11] If there are no clues to the contrary, readers can't help but assume the voice in the poem comes from someone who is identical with the poet.

Researchers have also looked at how readers react to a piece of literary text in comparison with, say, a few paragraphs about population growth. The act of reading—regardless of what you read—calls intensively on capacities for memory. However, expository texts tend to remind the research subjects of other facts. Literary texts, on the other hand, bring up personal memories of the reader as an actor rather than an observer. It seems as though personal information is called up to provide context for what we read.[12]

POSTMODERNISM seemed new and liberating to many writers in the second part of the twentieth century. But it created its own assumptions that need to be deconstructed—particularly, the assumption that there is a fundamental difference between "expert" and naive readers, that brains approach texts differently if they have been trained up, and that sophisticated readings are somehow different (and inherently better) than those performed by people who haven't taken university courses in literature.

In fact, human brains all tackle narrative from about the same starting point, regardless of whether we are expert readers or not. We all tend to love allusiveness, the pulling together of a vast network of information and experience in order to enjoy a work of words or images. Look at the density of allusion in an episode of the *South Park* cartoon series—you need to have so much information to get what's funny about it, everything from cinema history to current events. It makes "The Waste Land" look almost simple in comparison.

Postmodernism has also privileged a certain kind of opacity. Wild and whirling words that pointed to their own wordiness and not to an outer naive reality were somehow better—a moral judgement—than a plainly explicated poem because it was "good" for readers to be "challenged." (I remember my hackles rising when a reviewer referred to the "transparent afterword" in one of my books. "Transparent" was not intended as a compliment; I was not encouraging the reader to be challenged.)

I've heard quite a bit of sniffy snobbery about naive readers who give up because a poem doesn't mean anything to them and they can't figure the words out. That gets me royally inflamed at the sheer ego of writers who are pretending to efface themselves, forcing the reader to confront the poem alone, with no guidance. As if the poem were a dose of medicine to be administered by a faceless doctor in a white smock. "Here, take this. It's pure language and it's good for you. I'm not going to tell you what I was thinking when I wrote out the prescription—that might skew the results." When actually you're forced to go look up the author's biography and artistic theories to relate to the work at all.

"Postmodernism as a disguise for power," I mutter at the cat.

Then I have to tell myself to calm down. Sometimes I set up straw men and tilt at them. Poetry is more complex than that. The straw men come to life and wander around the field while I'm poking at them in a tin hat shouting, "Stand still."

Poetry *is* about "pure" language, and for some writers, that's the fun of it. For others of us, it's about "meaning," and for the best of poets it's about the intricate relationship as pure language and meaning meet and transfer their force to one another.

THE VIRTUE of challenging a reader is based on a conflicted view of the age-old dichotomy between emotion and rationality. Although artists would always say they were on the side of emotion and against the cold reductionism of logic, there has been, over the last century, an implicit moral superiority to art that deliberately turns away from the traditional ways human beings enjoy art. It takes the "higher" powers of judgement and analysis to appreciate an avant-garde work, along with a lot of training. On the other hand, a poem offering a straightforward engagement with emotion is now somehow automatically suspect, probably "superficial" rather than deep, and likely brushing up against the "sentimental."

However, that emotional-rational duality has been seriously shaken up by cognitive scientists like Antonio Damasio.[13] Now-classic studies of patients with brain damage show that the old picture of primitive emotion held in check by the new sophisticated functions of reason is basically out to lunch. Emotion arose in the long history of evolution as a way of organizing a response to incoming data and stamping certain combinations with significance. Without this organizing ability, rationality spins like a hamster wheel, as it does for the formerly successful stockbroker who, after a brain lesion impaired his capacity for a gut response, can no longer make good decisions. He can generate any amount of rational analysis to logically justify various courses of action, but he can no longer choose the best one.

Emotion is how we make rational choice. And in David Miall's analysis of how people read, he points out that our first response to literary works is emotional.[14] Whether readers are expert or not our first interest is not in interpreting a work. We do not immediately ask, "What does this text mean?" but rather, "What does it mean for me?"

Miall's own research supports the idea that narratives tend to be assembled in blocks he terms "episodes," each marked out by a distinctive shift in emotional response. This structure isn't merely a convention or a habit, he suggests: it's related to our brain's capacity to cope with a constantly new stimulus. "Feeling overcomes the limitations of working memory by providing a platform for registering the significance of the events as they unfold within an episode."[15] In other words, emotional engagement allows us to hold a story or a poem together in our heads.

WHAT COMES FIRST, the meaning or the sound?

A poem doesn't start until some phrase, some pattern of sound happens. I may have had the urge to write a poem; I may have a metaphorical template, ideas that I want to compare. But the poem never emerges from this egg without a small tap-tapping egg tooth to make the first chip.

"Broods on the nest of the east…": a phrase coalesces out of the yolk. I decide it has potential. There's the repetition / slant rhyme between "nest" and "east"—something of an obvious thump, but you need something obvious in a line to get attention. However, at least it's backed up by more subtle repetitions—the "n" of "nest" and "on," and the extra "s" in "broods." And the phrase also gives me a rhythmic pattern (DUM-da-da-DUM-da-da-DUM) that stands out slightly from the plod of normal speech. I find I don't want to lose it in a longer line. Perhaps a series of short, haiku-like stanzas?

Writing a poem is like picking a lock, poking a thin metal strip into a keyhole and manoeuvring it, listening intently for the clicks that tell you you're in.

MY OFFICE FLOOR has been stacked for a month in piles of poetry that tilt and slip from the vertical every time I walk through the door. I've been asked to judge a competition. My fellow jurors and I will draw a nice straight line between ten of the entries that will become real printed books, and the large remainder that will remain merely manuscripts. Wearily I scribble a single word on the cover of one of them and put it back on its slippery pile.

"Prosy."

It's one of the nastiest things a poet can say of another poet's work. "Prosy." The nose almost wrinkles in the very process of shaping the word. It's shorthand for saying there's little music, little sense of language *as* language.

Of course poetry and prose are both inherent in language, whatever we use it to say. The difference comes in at the level of intent—*why* we are saying something. Prose comes from the child's impulse to say, "I've got something to tell you." Poetry comes from the impulse that makes her chant, "Great green gobs of greasy grimy gopher guts."

Not that the two impulses can ever be totally separated. Part of the charm of the chant *is* the meaning, the frisson of grossness that appeals to a seven-year-old. And prose cannot eliminate musicality, rhythm—whether it is telling you a story, explaining a theorem or rendering a judgement.

However, the distinction between poetry and prose is real. It's why you can have great poetry that sacrifices syntactical meaning, but not great prose that does so. (*Finnegan's Wake* should perhaps be considered an almighty long poem.) And you can have a great novel (or scientific paper) that sacrifices the poetic potential of language, as much as that can ever be done given that rhythm and repetition are inherent in any string of words composed from a limited set of available sounds.

At any time in writing a line of poetry, I know if my intention in choosing a word or even a part of a word is related more to the requirements of poetry or prose. Should I, for example, decide on "brood*ing* on the nest of the east"? But no, that adds an extra syllable, throws off the rhythm. The extra "ing" interferes with the other sounds in the short line, which suddenly has too many sounds and becomes too close to the mélange of prose.

These micro-decisions remind me of the situation in chaos theory. When you plot solution after solution for certain common equations on the complex number plane, you see patterns emerging the way weather systems shift across a continent. The solutions tend to map out certain basins of attraction, where one behaviour or another prevails. But the interesting thing is that the boundaries of these basins of attraction never settle down to draw neat lines; at the edge, you may have one point that belongs with the solutions for basin A, while the next point calculated by the equation belongs to basin B. It's a constantly interwoven territory of different strands, points, slivers of this, flakes of that. Prose and poetry are just such basins of attraction, with infinitely complicated edges.

THERE'S A GOOD CHANCE that the manuscript I've just dissed as prosy will be on another juror's shortlist. My assessment may depend as much on aspects of narrative as on sound. I forgive other poems for relative prosiness if I find they are telling me something I want to hear, something I find unusual or interesting.

In fact, there's nothing like being part of a poetry jury to keep you humble. "Oh, gawd, not another writer going on about ecstatic contemplation of the prairie landscape," you mutter, turning with relief to the manuscript set on inner-city streets. But the juror from Toronto has seen quite enough down-and-out grunge, thank you, and finds the prairie book quite fresh.

So the components of story—setting, plot, character—are at least as important in engaging a reader in a poem as sound. Even the briefest lyric poem has a narrative moment that brought it into being—an orientation to where the poet is, what she is looking at.

> Oh, Hisagata
> where, over the waves, comes
> the song of cicada.[16]

We are near the fishing village of Hisagata with the ocean waves nearby and the sound of the insects—a juxtaposition of elements, something tiny managing to float over something immense. The poet hasn't given us any "I", fictional or otherwise, but he has put his hand on our sleeve and gestured. The reader is not puzzled about where she is or what she should be looking at.

WE NEED THIS HAND on our arm.

I've heard writers turn up their noses at the reader's "naive" compulsion to assume the "I" in a poem has something to do with the poet, to go looking for literal biography. "Haven't they ever heard of the fictional 'I'?" they sniff. Of course, readers know all about fiction. But a novel starts out "Call me Ishmael" and gives a few clues as to who the narrator is. A reader encountering a line of words in a poem needs something to orient the "I" wandering lonely as a cloud. We do need to know what the narrative moment is, the context for a poem. It's only human to want that, to know whether or not we are interested in making an emotional connection to the work.

I once heard the great Irish poet, Seamus Heaney, perform his sonnet sequence, *The Clearances*. He had, rather improbably,

arrived in Edmonton and was appearing at an auditorium in
the suburb of St. Albert. I had already encountered the poems
in one of his books. But when he prefaced them by briefly
mentioning they are an elegy for his mother, the poems suddenly
snapped into focus for me, and I realized how moving and
beautiful they are.

When I first encountered them on the page and admired
many of the lines and images, I had a vague idea that the
sequence was about a grandmother or perhaps several different
female relatives. Rereading the poems now, I wonder how I
could have got off on such a tangent. Perhaps if I'd attended
to the dedication more carefully ("In memoriam MKH, 1911–
1992"), I'd have guessed that the dates couldn't belong to a
grandmother. But that would have required a little mental
arithmetic and a knowledge of the poet's own birth date. So
I slurred carelessly over it.

I felt mildly exasperated with myself for being such a
slovenly reader. But I also felt mildly exasperated with
Heaney. Had the words "my mother" appeared anywhere in
the dedication or the nine-line poem introducing the sonnets
(where only an anonymous "she" is mentioned), I wouldn't
have had to go outside of the poems to get the relationship
that informs the whole sequence and holds it together.

We don't need more than the tiniest amount of framing, but
I think even this small amount of context is key because it
helps with the orienting response that is fundamental to
human brains. The orienting response originates in the brain
stem, where a cluster of ancient systems evolved to make an
animal pay attention to significant aspects of its environment.
It includes systems that keep us awake, alert and scanning the
environment for novel stimuli. As information comes in, it is

forwarded to other parts of the brain for further assessment of its importance. But the orienting response itself is instinctive and largely out of conscious control. We can't help being startled by a loud bang; nor can we help habituating to a recurring noise that then becomes invisible to our easily bored brain.

The flow of words on page after page of poetry can easily become a slur, even for a motivated reader. The very intensity of the language can be as habituating as the ack-ack-ack of rivet guns in a shipyard. That's why a little clarity regarding the narrative moment helps a reader so much. "All right, *that's* what we're building," she says to herself in relief.

And don't kid yourself—the hardest job a poet faces is the need to provide orienting information in a way that still lands on the poetry side of the prose-poetry equation. Especially when it's information you need to wrap into the first few lines.

"April opens the year with the first vowel, / opens it this year for my sixtieth," writes Mimi Khalvati—a graceful way of positioning the narrator's age and the time of year for a series of poems about childhood.[17]

Or, "I am poor brother Lippi, by your leave," writes Robert Browning, introducing his fourteenth-century artist in "Fra Lippo Lippi." It's an uncomplicated introduction, rendered sufficiently poetic by the line's pleasant rhythm and the repeated consonants of "p," "b," "r," and "l."

They make it look easy. It's not.

IN ADDITION to knowing where to start, a poet must also know how to stop the flow of narrative, which can become its own kind of habituation. We read "story" as much for what is going to happen as for what is happening now, and that can send us down the track of meaning while shutting out our attention to the sounds.

For me, one very interesting thing about Miall's research is that that we are literally stopped by poetic devices, such as alliteration, assonance and meter.[18] We pause for a few micro-seconds, unconsciously assessing/savouring them. Readers (whether advanced students of literature or relatively unskilled) are remarkably consistent about noticing such foregrounded components of a literary text, although our emotional reactions to them as individual readers may differ.

It is as though, paradoxically, the sounds of a poem function almost like rest spaces in a piece of music, a silence where we can linger over a phrase and integrate the ongoing progression of a score. In a poem, an unexpected combination of sounds gives us a moment to stop, assimilate, place the narrative in context while we attend to something different for a moment.

THE STRAIGHT LINE of my childhood has become, this autumn morning, the power line from the back alley to my house. Light runs rapidly back and forth on it like a squirrel. No amount of tightening could make it absolutely straight—it needs to sag, pulled by gravity. We exist, not in a single dimension defined by ends and opposites but rather in a more complex space. The line remains a useful concept for drawing triangles and other relationships, but it is outdated as a model for human thought. Science offers us another image—the field.

An electromagnetic field propagates through space in waves of light and magnetism. The electric field is oriented at right angles to the magnetic field but they travel in the same direction. Stop the field in its tracks at a detecting screen and you might say, "Oh, that's a poetry wavelength" (or prose, or lyric vs. narrative, or sound vs. meaning). But that registering sparkle is a function of the instruments you have laid out to detect it at that moment. Meanwhile, the field continues on its multidimensional way.

FIVE Symmetry

I was not impressed when my mother told me I was going to have a baby brother. It wasn't because I was jealous of losing some of my mother's attention, or because babies make a lot of noise. But I *was* seriously annoyed that it would break up the symmetry of our dinner table.

For as long as my conscious memory extended back, there had been four sides to the table and one person to each side. Two daughters mirrored each other on two sides, parents were reflected on the other two. This arrangement of my world had stability and balance. Where would this new child sit? How could you fit five people around a four-sided table? It would inevitably become heavier along one edge, out of whack, out of kilter, for every dinner, forever.

I'm not sure why an eight-year-old would have been so fixated on the symmetry of family around a chrome-legged Arborite kitchen table. Perhaps it was somehow a metaphor for a deeper concern about the family dynamics. Perhaps it was the sign of an overly prissy personality. But I have a very clear recollection of worrying how the now-asymmetric table would *look*.

THE HUMAN VISUAL SYSTEM is extraordinarily fond of symmetry, especially the bilateral type where the left side of a face is reflected in the right, or one hand mirrors the other. We detect such balanced reflected patterns immediately and unconsciously. Symmetry perception is robust—we extract symmetrical patterns from data coming at us whether we consciously intend to or not. Yet this facility isn't rooted in relatively passive physical systems, such as the fact that neurons in our visual cortex are wired up in geometrical arrays (though to some extent they are). Recognizing symmetry is an active process, and we are also gifted at easily recognizing the almost-symmetrical. We see both the generalized pattern of reflection and the tiniest of deviations from it.

Humans tend to like the appearance of bilateral symmetry very much. When you take photographs of human faces and tidy up the tiny asymmetries from one side to the other—the slight misalignments at the corners of eyes, angle of eyebrow, curve of lip—we judge the more symmetric versions to be more attractive. I've looked at some of these test pictures myself, and in spite of my conscious prejudice against prettification and bland artificiality, I can't help it. It's almost impossible to articulate why, but the more symmetric versions of faces look more pleasant, more appealing. Other animals (which are not particularly indoctrinated by cultural expectations) also seem to prefer symmetry in their fellows.[1]

Researchers have completed years of experiments, research subjects peering at dots and lines, brain scans and computational analysis, but we still don't completely understand why human beings have this innate ability to detect and enjoy symmetry.[2] Is it simply because it's easier and more efficient to process similar signals from different parts of the visual field, a kind of mental economy? Is it because we are members of the great family of Bilateria, along with almost all other animals on the planet, with our brains and bodies aligned along an axis of reflection? After all, there are big risks and rewards associated with the other animals that turn up at nature's table with us. It would make sense to evolve systems for picking them out of the surrounding vegetation.

Still, regardless of how and why we handle this complicated task of visual analysis, we do it easily. Our pleasure in symmetry makes it one of the major features of art in all cultures, times, forms.

HOWEVER, we're not *that* good at recognizing maximum symmetry.

Imagine you and I are standing on a sheet of white paper that extends to infinity in all directions. Nothing on it; no knives and forks, no dots or curves. "Absolute sameness," we think. "Nothing symmetrical here." Yet, to a mathematician, this space of apparent nothing has infinite symmetry. You could put an axis of reflection anywhere—not just down the middle of a face or the centre of a table—and the original would be indistinguishable from the reflection. You can rotate the plane or shift a patch of it in any direction and nothing will seem to change.

To the mathematician who has joined us to point out the beautiful sameness of the Euclidean plane, a symmetry is something that happens but does not happen—an operation that leaves the end product indistinguishable from what you started with, a curious combination of transformation and stasis. With her eyes focused off in the distance, she quotes the great geometer Donald Coxeter: "All of mathematics is the study of symmetry, or how to change a thing without really changing it. It is symmetry, then, in its various forms which underlies the orderliness, laws and rationality of the universe, and thereby also the language of mathematics."[3]

While the mathematician is rapt in this infinite reverie, you crouch and print a single letter on the white surface. Let it be that highly symmetrical letter "I," the axis of its own reflection. I look at it from the other end and nod. It looks the same from there. "Isn't that terrifically symmetrical," we tell each other.

But the mathematician slaps her forehead and groans. The infinite symmetry of the plane is now broken. From now on, there is a place on it that is distinguishable from the rest. If you rotate the plane around a point, our letter "I" will show up at different locations from where it is now; you'll have to rotate it right through 360 degrees before things look the same again. If you put down a mirror-axis on a patch of the plane, you'll now realize, "Oh, that half is different from this one," just as your left hand differs from your right.

"You've gone and messed it all up," says the mathematician, contemplating this infant line. We don't sympathize. "You've got to start somewhere," we tell her. "We can't stand around her forever looking at nothing. Something new has to happen. Let's make another one."

DIFFERENT MODES of perception are biased in how they detect symmetry.[4] Our visual systems handle reflection superbly but are less clever at noticing translation symmetries, the kind of transformation that picks up a pattern and shifts it to another part of the plane. When researchers have us peer at configurations of dots and lines, we're less able to detect that the same pattern has been repeated (but not flipped into its mirror image) on the other side of our visual field.

However, when it comes to hearing, our detection capacities are reversed. We don't easily detect that notes have been mirrored, as in when an identical sequence of notes has been repeated backward. Nor do we quickly hear palindromic words like "rats" and "star" as being made up of the same sounds forward and backward. But our ears do recognize *translation* symmetry—a chunk of sound that has been shifted like a tile on a floor—very well. This bias applies in both music and poetry, which is not surprising since poetry is first of all an oral art form and has often been compared, rather vaguely, to music. "That line is so musical," we say, meaning generally that it sounds pleasant in a rhythmic way.

"Shall I compare thee to a summer's day..." The ta-TUM, ta-TUM of that iambic meter is the heartbeat of English poetry. Music and poetry share the same tension between a regular repeated rhythm plodding away in the background and the phrasing of a real combination of words or musical ideas that floats over top of it.[5] In music, the ideal pattern is marked off by bars, within which notes must add up to a consistent sum. In poetic meter, the patterns of stress in different types of metrical foot have a similar kind of time-keeping function. You're supposed to get the same number of stresses in each one.

Actually, even more than the exact number of stressed syllables, the idea of a poetic foot is to have roughly the same amount of *energy* in each one. Musicologist Robert Jourdain notes that we process the tempo of music and of language in similar ways and that we unconsciously slow the rhythm of speech at the beginning and end of phrases. "It's not the velocity of movement that's constant, but rather the level of difficulty of movement. If you were to set a metronome to an average speed and then move to it in lock-step, you'd drag your feet across the centre of rooms but spin out at corners..."[6]

A REGULAR BEAT, whether it's a quick-march in music or an iambic pentameter poem, has very restricted symmetry. Like one of those patterns crimped around the rim of a pot, meter has limited dimensions. It's a one-track line in time.[7] A better example of how music and poetry use translation symmetry lies in the similarities between melody and rhyme. Both repeat patterns in several dimensions at once, pieces of soundscape shifted sideways in time. Both are clusters of sounds recurring in the same order, and especially they depend on recurring patterns of stress.

Melody is crucially dependent on rhythm and stress. What distinguishes a melody is its pattern of duration and accentuation, and notes that fall on downbeats or other important rhythmic junctures usually become the tune's most recognizable ones.[8] Rhythm is so important to melody that we can identify familiar tunes when their rhythm is tapped out on a single pitch—although we do not do as well when the tune's tones are used but they are all made the same length.[9]

Rhyme is just as dependent on stress patterns. This is why "entered" and "interred" are not rhymes, in spite of the fact that they group almost exactly the same sounds in the same order. However, we happily accept Jack and Jill's pairing of "water" and "after" as a satisfactory duplication of sounds.

We learn to recognize and repeat melodies and rhymes in very similar ways. One of the first things a brain perceives about melody is its overall contour—its rising and falling pattern. Babies as young as six months will notice a change in melodic contour, but they won't register that a melody has been shifted up or down in pitch. However, to *sing* a melody takes children longer.[10] As Robert Jourdain points out, children must learn a great deal about their culture's tonal systems and how the continuous slide of sound up and down gets chopped into notes before they can truly sing. Until children are three or four, they'll identify different melodies as the same so long as their contours match roughly. However, real melody needs us to replicate tones of fixed pitch and duration.

We go through a similar process in order to learn the language sounds, the consonants and vowels that will eventually allow us to rhyme. A wide range of sounds approximates "p" and a child must learn which variants count as meaningful in his own language. These meaningful phonemes involve complicated positioning of tongue, lips and larynx. For instance "p" is formed, not just by closing and opening the lips, but also by shutting off the flow of air in the larynx so that it doesn't go over the vocal cords. If you forget to do that, your "p" turns into a voiced consonant, "b." If you forget to keep your teeth away from your bottom lip when you puff, your "p" turns into a "ff"; forget to turn off the vocal cords and "f" turns into "v." We can hear such sounds morphing one into another in words like *pater–vater–father* as languages drift and glide apart.

Getting sounds to match is the essence of learning both music and language, and we begin to be able do so around the same age. The simpler rhymes and melodies of childhood reward us with delight in being able to recognize and reproduce sounds that are the same. They become the basis for a lifelong source of pleasure. A sequence of sounds, a fragment of repeated melody, registers quickly and clearly, even when repeated after a considerable length of time, and we welcome them back when they recur. Our enjoyment comes from identifying the recurrence of a complex structure, from recognizing how it has been translated through time.

After we've got the basics mastered, we find melody and rhyme easy. Untrained adults discern melodic contour almost as well as trained musicians.[11] We often don't value what comes easily to us and dismiss it as trivial, but the things that come easily to human beings are often the most deeply complicated functions to reproduce. "Cat, hat, mat," the kindergarten kids chant, and it seems simple, unsophisticated. But even "cat" and "hat" require sophisticated mental processing to recognize what phonemes are present in the rhyme words and how they interact.

The four-year-old learning a nursery rhyme is learning how her language is constructed of syllables. She's learning the subtle variations of sounds within words and the way different consonants peg vowels down in different ways—the "a" of "cat" and "pat" is not the same as the "a" of "car" and "far." Most important, she's learning how sound adjusts itself to rhythm. When she recites, "Peter, Peter pumpkin eater / had a wife and couldn't keep her," she finds that "eater" and "keep her" have something in common that overrules the difference between "-t-" and "-p h-."

AS OUR NEW BROTHER grew out of his high chair, my sister and I were occasionally assigned responsibility for babysitting. Mostly this involved seeing how high a bridge we could build with the interlocking rail sections of his plastic train set, using books pulled from the bookshelf as support pillars. The bookshelf had a random assortment useful for bridge-building, including three dark-blue volumes of *The Modern Painter and Decorator* from which my father had learned his trade; a set of ten volumes of *Newnes Book of Knowledge* (an illustrated encyclopedia that we called "The Big Books"); Hans Christian Andersen and *What Katy Did* and *Blackie's Omnibus of Brownie Stories*. It also held a scattering of Little Golden Books—tough-covered and colourful—which my mother bought at the grocery store for 25 cents apiece. Sometimes, when we'd used up all the thicker books and the plastic rail sections, we'd sit down and read one of the Little Golden Books to David. One was about colours and on the page about purple it had a delightful piece of text that locked itself in my brain:

Purple as pansies, purple as plums,
Purple as shadows on late afternoons.

I read that with almost a shudder of pleasure. For the first time I recognized that words as different as "plums" and "afternoons" could be used as rhymes.

WE PERCEIVE both rhyme and melody as a grouping of sounds that sticks together. The ability to assemble the world's contents into discernible wholes is inherent to all kinds of cognition.[12] Our brains group patterns according to Gestalt principles first observed early in the twentieth century. One such principle is completeness: we prefer a completed pattern, a resolution in time. In music, we want melodies to conclude with a cadence (a note or chord that brings us back to the harmonic trend line). In rhyme, the defining feature is the *end* of the sound cluster—the onset sounds are different, but they have to come back satisfactorily to the same place that the first word of the rhyme partnership laid down.

"A melody is an auditory object that maintains its identity in spite of transformations," writes Daniel Levitin.[13] You can shift a tune up or down in pitch, but our brains easily recognize it as the same. We like such manipulations—exact repetition doesn't please us. We want things twisted up a bit. Any pop song composer knows to use a deceptive cadence in the first few recurrences of a melody, then come to the expected cadence for the final stanza.[14] And, in poetry, the essential nature of rhyme is a combination that keeps some components the same while changing others.

HERE IS THAT chrome-legged 1960s table again, now in a different kitchen with one end jammed up against the wall so that its capacity for symmetry has changed. I set the cutlery for five—two sets on one side, two on the other, and Mum's at the end closest to the stove where there will be smooth, tinned, tomato soup for our Sunday lunch. A kind of balance has been restored.

"Say a poem, Daddy. Say 'Our Faither's Faither's Faither,'" we clamour. My father launches into his rhyming comic tale, our favourite, about a working man who gets hold of a garbled version of Darwin's theory of evolution and takes the family up to the Glasgow Zoo,

> *...just to see for mysel'*
> *all my faither's cousins*
> *who hadnae done sae well.*

There are fights on the bus, bored zoo attendants in skippet hats, whining children and aggravated wives. The stanza

> *We staggered round that blooming zoo.*
> *The bairn was greetin' for a kangaroo.*
> *The wife, she wasnae saying much,*
> *but I could tell by the look in her eye*
> *she'd have had my guts for garters*
> *if there was naebody passing by*

always dissolved me into a Jell-O of laughter. He'd get to the concluding couplet, where the father is contemplating the monkeys:

And you know the thought that struck me?
Aye, it's probably struck you too.
My faither's faither's faither

and we'd all shout together:

would have been better aff in a zoo!

Rhyme was the first feature of poetry my dad had discovered. As a small boy, some chance remark of his made his mother turn around. "Why, Wullie, that's a wee rhyme you made there." Catching the erratic attention of his mother was such a huge reward that he started making up more and more rhymes to bring her. This was one of the formative stories of his life, one that he told again and again at the end of it, when the memory of other episodes was long gone. And the rhymes of his poems also stayed with him even after all the stories left. He could recite scraps of poetry to the nurses, as his mother had done in her failing years.

In between, as an energetic, competent, maddening and lovable man, he composed rhyme after rhyme, recited them around the kitchen table, made us laugh until we cried.

RHYME has been out of fashion in literary poetry for more than half a century, its manipulation seen only as a clever craft. This is particularly true of end rhyme, the matching sounds at the end of metrically similar lines that define traditional forms from ballad to *terza rima*. Conventional wisdom maintains that English is not a naturally rhyming language, unlike French and other Romance tongues. Instead, the early English poetic forms were based on alliteration. What John Milton called "the troublesome and modern bondage of rhyming" was a fancy fashion imported from the Continent by poets like Chaucer.[15] Over centuries of flogging the combinations of love/above or night/flight/light, we've wrung them out.

Certainly I realized it was out of fashion when I first tried publishing poems. I would disguise my sonnets by changing their line breaks, turning the end rhymes into internal ones so that the translation symmetry of sound was still there without dangling conspicuously. Sometimes this manipulation may even have improved the poems, but it was done more in the spirit of changing a hemline so I didn't look dated in the mirror.

This conventional wisdom overlooks the enthusiasm with which we adopted rhyme for serious poetry in English. A fashion that lasts for six centuries is something more fundamental than a culturally acquired habit. That old enthusiasm has sprung back vigorously in the polysyllabic rhyming of rap, which has migrated into languages as varied as Danish and Hindi—languages that are no more or less "natural" for rhyme than English is. I would argue that a stressed language like English actually works very well for rhyme because stress profile is so much a part of what makes a rhyme work. As Robert Pinsky notes, rhyme is a matter of degree, not an on–off toggle.[16] When Eminem can rhyme "Munchausen syndrome" with "public housing system," you can see how flexible the process is.

And the idea that rhyme combinations can be used up is, when you think of it, rather silly. It's like saying words like "dark" or "table" have been exhausted because poets have employed them for centuries. Language constantly renews itself because words constantly have to talk about new things, new experiences, in new configurations.

I WAS LOOKING FOR a rhyme for "microwave." This was the end stage of a long project, *The Office Tower Tales*, in which I had been using a five-line stanza with end rhyme. I had been brave enough to let my hemline creep up to where it would attract a little attention, though not to the head-swivelling levels of complete metrical regularity. I felt I needed something clearly poetic to carry my long comic narrative of women telling each other stories on their coffee breaks in the food court of an office tower. As well, the rhymed frame narrative set it off from the tales themselves, which were mostly in more familiar free verse. But this bright idea had meant months— years—of struggling with the questions of rhyme. What rhymes and what doesn't? What makes some word combinations register as similar enough?

At this point I had finished the end of the project but still needed to go back to write the first page or two. In these opening stanzas, I wanted to set the tone for the whole project, to create (for those readers who might care) a kind of echo of the opening lines of Chaucer's *Canterbury Tales* ("Whan that Aprille with his shoores soote..."), which had been my original inspiration for the whole project. And I had to make sure the stanza form was sufficiently clear for readers to sense it. It had taken a while for me to figure out what constraints I would work under, and some of the rhymes in the earlier part of the frame narrative were quite loose. I wanted the pattern to be fairly clear here, where readers would first encounter it.

I also wanted to work with a particular visual image of Earth going around the sun the way a coffee cup goes around in a microwave when it's off-centre, both planet and cup being warmed by radiant energy. So I started going through the little exercise my father taught me long ago—take each letter of the alphabet in turn and try out the sounds.

"What rhymes with wave?...bave, cave, dave, fave, gave..." Then more carefully through the other initial sound combinations: *brave...crave...architrave...* "Crave" caught my ear quickly. It fitted with the sense of urgency, the desire for spring that drives us to ditch our boots and turn our faces to the strengthening sun. So the lines became:

> *...The planet's orbit spins into the phase*
> *of northern warmth*
>
> *like a coffee mug revolving*
> *slightly off-centre in a microwave,*
> *absorbing heat into its porcelain bones–*
> *morning sun the caffeine we all crave.*

I read this over and considered it good, continued on with my chain of words. But afterward, going back over the prologue (and especially reading it aloud), I became less satisfied. In fact, I don't hear "microwave" as a really good predictor for "crave." The problem is rooted in rhythm. When we say, "Oh, just warm it up in the microwave," we tend to give the word a slightly dactylic pattern—TUM-ta-ta—so that the long "a" of the "wave" syllable is downplayed compared with its definitely prominent nature in "crave." The sounds that are

most prominent are the ones in the first, "micro." The long
"a" surges back if you say, "Oh, just microwave it," because the
following "it" changes the rhythm of the phrase to a trochaic
one—TUM-ta TUM-ta. "Microwave it" and "crave it" make
an excellent rhyme. However, the rhythm of "we all crave" is
spondaic; all the words have to bear about the same amount of
stress. TUM TUM TUM.

Would it help if I could somehow get the sounds of "icro"
into the line with crave? (Cryo-? Ichor? I cry? Yikes!) Maybe if
I tried rearranging the lines so that some other words became
the focus of rhyme. (Bones? Cup?) If I added a hyphen to the
compound word (micro-wave), would it help people read it as
three stressed syllables: MI-CRO-WAVE? I spent hours on that
aggravating verse. It didn't help that I'd already written the
remaining stanzas needed for the prologue, and if I mucked
around with this one so that it changed the lines immediately
following, the changes could reverberate for several pages.

I was still trying to find an alternative right up until the
galley proof stage. Then, I said at last, "Oh, the hell with it."
Sometimes you have to accept the fact that the symmetry is
less than perfect, that there's an extra place setting on one
side, and hope the universe recognizes your good intentions.
Otherwise you'll never get on with things.

IN FACT, that's what the universe had to do: break the initial
perfect symmetry in order to get on with things. During the
initial infinitesimal moment, everything was the same. It was
not just the sameness of the Euclidean plane where our mathematician was standing a while back; it was sameness in every
possible dimension. There was no separation of matter and
energy, forces and particles, just a completely homogeneous

soup no more than the size of the Planck length—a distance even smaller compared to a proton than the proton is compared with Mount Everest. But this phase of complete symmetry was unstable. The newborn universe ripped itself to pieces to cope with the enormous pressure.

"Although we physicists value and admire symmetry, we still have to find a connection between a symmetric theory and an asymmetric world," writes physicist Lisa Randall.[17] The particular story of our universe is one of symmetry-breaking, of things becoming more and more unlike.

As the new universe expanded and cooled, four physical forces that govern our world today cracked out of it, rather as water goes through a phase change when its molecules exchange the greater symmetry of fluid for the more limited symmetry of ice crystals.[18] Gravity separated from the strong force (that binds quarks into the particles of matter), the weak force (that helps assemble protons and neutrons and electrons into atoms) and electromagnetism (the energy of light). A zoo of different particles was now zooming around— hadrons, electrons, photons, muons and neutrinos, all of them in assorted flavours and colours—each of them an outward sign of the forces with which it interacts. Instead of the homogenous soup, we are on our way to stars.

But when you reverse the symmetry-breaking and return even a tiny part of the universe—a few atoms in a supercollider— closer to the hot high-energy conditions of the early universe, you can see the forces recombine themselves. So far we've only been able to do that with the electromagnetic and weak forces, but with new and ever-more-super-colliders we may be able to unite the others, as well.

LET US BE FANCIFUL and think of forces in poetry as being similarly connected at the root even if they seem to present themselves as separately observable. In particular, rhythm and rhyme may seem to be different, but they interlock. In traditional forms of poetry, end rhyme is the cymbal crash at the regular intervals of rhythm.

What rhyme and rhythm share at heart is expectation—the anticipated recurrence of a symmetrical pattern. They get our attention. This expectation is most noticeable with rhythm; a regular beat sets up a train of pulses and the brain continues to anticipate them with remarkable accuracy even when individual pulses disappear into silence. However, if the train of pulses is not eventually renewed, our anticipation of them wanes.

"Psychologically, pulse constitutes a renewal of perception, a reestablishment of attention," writes Robert Jourdain.[19] Rhyme does the same kind of thing, but in a different direction. It kicks you back to hear again something that you heard before—something you may not even have noticed the first time. While rhythm generally goes one way in time, rhyme has a funny way of operating backward and forward.

It does this by delaying the processes in the brain that strip away auditory information when incoming sound is handed off to the semantic modules that assign it to an abstract category (like "p" vs. "b"). We lose the actual puff or pop quickly from consciousness and focus on assembling meaning. Rhyme delays this process of stripping out the sensory data, allowing us to hold it in a kind of echo chamber of the brain, the auditory short-term memory.[20] Reuven Tsur, one of the earliest scholars to apply cognitive theory to literary studies, points out that in some circumstances, rhyme will reverberate more intensely and longer than most other aspects of poetic language. The brain perceives rhyming units as being closely knit together even if they are relatively far apart, so that rhyme spreads "a kind of sensory net over a considerable region of a poem."[21]

Rhythm and rhyme are both forces that establish expectation and deliver on it; they get our attention. Nowhere is this interlocking clearer than in rap, which is defined by its four-beat lines and a hail of multisyllabic rhyme—internal, end-stopped, in the pocket or off-beat. In literary poetry, the energy has expanded off into the cooler regions. Rhyme, like the weak force, may almost disappear from day-to-day observation, only to show up sporadically like an act of random beta decay. The original pulse of rhythm has become muted, like the cosmic background radiation that is today's faint remnant radiation of the Big Bang's massively uniform pulse.

In fact, poetry has cooled off the initial thump of repeated meter so much that it has traded its aural symmetries and transformations for visual ones. This process has been going on since the beginning of writing and more and more quickly since poetry became a printed thing. Concrete (or emblem) poems have been shaped in a symbolic ways since the seventeenth century. For example, the first stanza of seventeenth-century George Herbert's "Easter Wings":

> *Lord, Who createdst man in wealth and store,*
> *Though foolishly he lost the same,*
> *Decaying more and more,*
> *Till he became*
> *Most poore:*

> *With Thee*
> *O let me rise,*
> *As larks, harmoniously,*
> *And sing this day Thy victories:*
> *Then shall the fall further the flight in me.*

Such visual patterning was taken up by everyone from Lewis Carroll (in his poem from Wonderland, "The Mouse's Tale") to the Dadaists of the early twentieth century, and in the process lost much of the rhyme and meter that Herbert was using. Today, it's one of the things kids love most to do in a poetry workshop—make their poems in the shape of a tear or a baseball bat. The computer technologies that allow such experimentation are more and more available and flexible. But the more we focus on what the poem looks like, the less we hear how it sounds. Even poems that are not put into shapes reflecting their subjects are defined by their visual symmetries on a page. A poem of three-line stanzas, heavily enjambed, can be balanced and crafted on the page but give you nothing as a symmetrical unit to *hear*.

However, there's still an old unity linking the auditory and visual in the apparent struggle of free verse. When you're just a little experienced in reading contemporary poetry, you realize visual information can convey an auditory pattern. In the 1970s poet Charles Olson could write of a still-newish technology: "It is the advantage of the typewriter that, due to its rigidity and its space precisions, it can, for a poet, indicate exactly the breath, the pauses, the suspensions even of syllables, the juxtapositions even of parts of phrases, which he intends. For the first time the poet has the stave and the bar a musician has had."[22] However, the systems of translating sound into pattern on a page have tended to be idiosyncratic and not particularly well defined to the general reader. Take a couple of Olson's own lines:

> *They buried their dead in a sitting posture*
> *serpent   came   razor   ray of the sun*

How long *do* you pause between those separated words? As long as a comma? As long as a semicolon? How does the pause differ in length/quality from the line break after "posture"? A reader unfamiliar with Olson's particular set of conventions has little idea how to make this visual pattern into sound.

RAP ARTISTS don't face this particular hassle of translating between auditory and visual. In performance, the two kinds of experience are still an unbroken symmetry. In *Book of Rhymes: The Poetics of Hip Hop*, Adam Bradley points out the close connection between the artist as performer and the artist as wordsmith. These are skills that have drifted apart since the days of Greek poetry competitions and itinerant Celtic bards, but rap restores the connection—and then some. The free-styling MC is "not simply a performer but something more: an artist conceiving the lyrics before our very eyes," setting a standard of white-hot creation.[23] Of course, actual performance may be a little further from the Big Bang. As Bradley points out, rappers may have a distaste for repeating "cover tunes" but they compose their pieces carefully, perform them over again at concerts and even occasionally use ghostwriters. But the ideal is a unique, unrepeatable experience, inseparable from the creator.

The success of rap is not surprising. It wasn't just a response to social conditions faced by black youth in the boroughs of New York City. It isn't due to the MC's unfair advantage in having a beatbox or turntables to get a head-banging rhythm going. It's not simply a fickle fashion. Listeners responded so strongly to the rhyme, the wordplay, the rhythm, because there was a vacuum to be filled. Humans have always played with language, "making it special" in the phrase that Ellen Dissanayake uses to denote one of the main markers of art.[24] Part of the popular reaction to free verse and contemporary poetry was that it didn't seem to make language special enough—its sonic symmetries were too subtle to break out of the background.

From my father, I learned the joy of rhythmic language and rhyme. But hip hop wasn't exactly an option for a girl in Scarborough at the end of the 1960s, so I learned to write the poetry of my time; I write for the page. I'm not sorry about this. I like that the encounter with printed poetry can be replicable. I carry out my symmetry operations of transforming sound to visual image in the hope that readers can reverse the process. And I like having the option of separating from my work, the fact that I can send the words away from myself as a form of radiation that may take them farther than I want to go myself.

A YOUNG BLACK MAN is standing at a battered podium with a buzzy microphone in a lobby of the downtown library. It's the middle of the Edmonton poetry festival and here at "Poetry Central," people are invited to stop by and read a favourite poem aloud. Roylin begins to recite in a voice shaped by his Louisiana childhood, a long way from the northern plains of

Canada. His choice is somewhat surprising—that old standby, *The Shooting of Dan McGrew*, by Robert Service.

It takes me a few moments to realize, "He's doing it straight!"

Roylin wasn't camping it up for laughs. He was treating the poem as a story of betrayal and loss. "A half-dead thing in a stark, dead world, clean mad for the muck called gold." Perhaps because he knew the more contemporary traditions of rap, where rhyme can be used to tackle all sorts of subjects, he wasn't compelled to treat the tale as funny.

Now why should I have been assuming humour? If Roylin had decided to recite a poem by Keats or Shelley, I might have been as surprised by the selection but I wouldn't have immediately expected humour because it was a rhyming poem. Perhaps it's just that my mind associates Robert Service with his best-known poem, *The Cremation of Sam Magee*, which is definitely comic-macabre. However, I think the assumption of comedy really has to do with broader assumptions about English poetry.

Dan McGrew is written in language that is essentially modern in its sentence structure and vocabulary. It has none of the inversions and archaisms that had been preserved in Victorian poetry: "The cloud whose bosom, cygnet-soft, / A couch for nuptial Juno seems."[25] But its rhyming couplets hit a rhythmic stride almost instantly:

> *A bunch of the boys were whooping it up in the*
> *Malamute saloon;*
> *The kid that handles the music-box was hitting a*
> *jag-time tune;*
> *Back of the bar, in a solo game, sat Dangerous*
> *Dan McGrew...*

There's nothing inherently hilarious about that, but it includes a lot of dactylic feet (a rhythm that goes TUM-ta-ta). Dactylic rhymes (the kind that match "dimity" and "magnanimity") are relatively rare in English, but they have become almost entirely locked to humorous poetry since the days of Lord Byron:

> *He'd learn'd the arts of riding, fencing, gunnery,*
> *And how to scale a fortress—or a nunnery.*

Service doesn't use dactylic rhymes at all in *Dan McGrew*—they are all good, thumping, one-stress rhymes like lash/crash and hear/dear. But that dactylic pattern canters through his lines: "BUNCH of the...WHOOping it UP in the...HANDles the...HITting a...BACK of the..." The comic expectation has been transferred from a particular rhyme stress pattern to that rhythm in general.

Robert Service published *Dan McGrew* in 1907; only a couple of years later, T.S. Eliot was composing a vastly different poem. It was also written in modern vocabulary and syntax and still used a fair bit of rhyme, but the metrical patterns had been loosened dramatically:

> *Let us go then, you and I,*
> *When the evening is spread out against the sky*
> *Like a patient etherized upon a table;*

"The Love Song of J. Alfred Prufrock" became one of the signature poems of the twentieth century; Service's poem was popular but definitely downmarket.

Of course, Eliot deserves to be considered a much greater poet. Robert Service's works have a limited range that swings between comedy and melodrama. He is frequently hokey and this is the real reason he is enjoyed but not revered. However, as the decades passed, there came to be a general sense that any poet who used meter and rhyme had to be writing comic verse, not because of their concerns or talents, but just because of the form they were using. Rhyme could be clever in the hands of Ogden Nash or Dorothy Parker, but rhyme—especially in conjunction with lighter rhythms—was not for serious subjects. Young poets modelled their work on the forms of Eliot, not Service. By the 1970s, it would have been hard to imagine a book of rhyming poetry by a young poet in Canada even being considered for publication.

Reuven Tsur points out there's no particular cognitive reason that such highly metrical rhyme patterns must be comic. However, there is something special about how our brains process the dactylic meter that gives it a particular emotive power.[26] It's related to markedness, how readily noticeable we find it.

This is partly because of the dactylic pattern's relative scarcity in English. A run of truly unstressed syllables is hard to come by. We say "in a PINCH." But if that phrase were preceded by another unstressed syllable ("use a staple in a pinch"), we tend to return a bit of weight to one of them ("USE a STAple IN a PINCH"). As Robert Pinsky points out, when we're reading an iambic line, (ta-TUM-ta-TUM-ta-TUM), our internal rhythm detector is not assessing how strongly a syllable is stressed compared to the rest of a line, but relative to the ones that are immediately next to it.

But more than the fact that we tend to shy away from multiple unstressed syllables, the dactylic measure does something else to draw attention to itself. Brains handle the world by grouping its elements. Unstressed syllables have to be grouped with stressed ones—either forward, with the stressed one that follows them (like "in a PINCH"), or backward, with the stressed syllable that came just before (like "DIM-i-ty"). Of these two possibilities, we tend to notice backward groupings more intensely.

As well, we do amazingly complicated and subtle things with our voices when we are stressing a syllable. We can accent it by saying it louder or longer or at a different pitch. In a beginning-accented group like "dimity," we say "dim" just a little louder. In an end-accented group of syllables, we draw the last one out slightly.

So the TUM-ta-ta combination is unusual, tends to register as a grouping because of its order and is just slightly louder. No wonder we notice it as the most marked of rhythms in English poetry. However, all rhythmic patterns that poets use are more marked than those of regular speech—which means in turn they have a capacity for emotional resonance.

In Antonio Damasio's somatic marker hypothesis, pleasant or unpleasant stimuli felt in the body are stamped with an emotional tag by the brain, enabling quick judgements when creatures don't have the time or capacity to rationalize a response.[27] These brain-gut connections are practiced over and over again during daily life and form the substratum for conscious decision-making. Robert Jourdain takes this a step further when he discusses our reaction to music: "What's enticing about [the somatic marker] theory is that it suggests that we've a long-evolved neurology for the explicit muscular representation of emotion." Movement is any nervous system's raison d'être, writes Jourdain.[28] In bodies that are expert in timing and moulding temporal shapes, rhyme and meter are noticeable, easy to tag with an emotional response.

It's no coincidence that rhyming poetry fell into disfavour at around the same time as a deep distrust of sentiment in poetry emerged. It is paradoxical that an art form so deeply committed to expressing emotion should become so suspicious of actively doing so. To some extent this may have been a necessary balance to the excesses of Victorian literature, which cranked up emotion and considered that a poem that could make you weep was a better than one that didn't. However, the correction has been intense. I remember well the sneer associated with sentimentality in the university English classes of the early 70s. William Carlos Williams's celebrated red wheelbarrow poem was written after a night at the bedside of a desperately sick child, but to directly mention the child and describe that situation would have been to court pathos. Such a poem would have been fit only for greeting cards or the poor sods who didn't know any better than to like Robert Service.

"Och, I'm only a makar," my father would say, meaning that he was simply a versifier, not a real poet. Even though putting words into a pattern and performing them gave him more pleasure than just about anything else in life, even though his poems could make people laugh and cry, he knew he didn't count as a poet by the measures of his time.

HERE IS the kitchen table again.

"What's for supper?" I ask my sister, and she responds in lugubrious tones appropriate for reciting Edgar Allen Poe, "The white dinner."

The white dinner consists of rolls of poached sole, boiled cauliflower and potatoes, all draped in a pale white sauce. My mother made it as a special, slightly fussier-than-usual meal, unaware its unbroken pallidity sank our hearts.

How often we do not realize things are special.

The five-fold symmetry of our family table, the pentagonal pattern, is found in nature very often. My mother's African violets on the high windowsill, the buttercup I held under my chin, quietly express it. However, five is a rather special number from the point of view of symmetry.

At first glance it seems less tractable than others. You cannot tile a plane continuously with pentagons as you can with other regular polygons which will fit together with satisfying snugness. We could stand on that infinite, white, Euclidean page again and see it covered with identical triangles forever. (Or with the squares like the kitchen linoleum, or the white hexagons that I stared at on the bathroom floor.) However, try tiling a surface with pentagons, and you soon discover that you have to come up with a second shape to fill up the corners.

Since Johannes Kepler, geometers have been trying to come up with regular shapes that will continuously cover the plane. One of the questions that interested them was whether all such tilings must be periodic: would they have to repeat themselves, so that you could slide a part of the plane over another part and have the two patterns identical? Translation symmetry, in other words.

In 1974 Roger Penrose identified two shapes that will fit together in an intriguing way. He called them "kites and darts," and they are essentially two pieces of a rhomboid that have

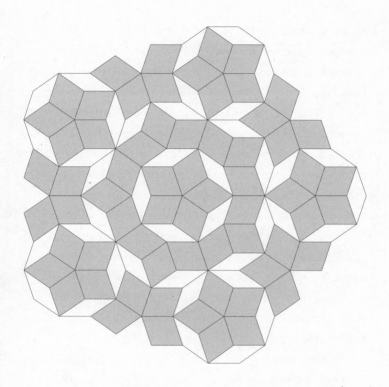

been cut apart in a particular way.[29] Put them back together as they started and they can tile the plane in a very ordinary, periodic, repetitive way. The resulting pattern has a lot of two-fold and four-fold symmetry. However, flip one of pieces over into its mirror image and you can assemble the two shapes very differently into a pattern that will never repeat itself, ever. No matter how far across our infinite white plane you crawl on your hands and knees with an X-Acto knife, you'll never be able to find a parallelogram that you can cut out and repeat, repeat, repeat to fill the plane.

What does emerge are recurring shapes with *five-fold* symmetry—pentagons and stars and five-spoked wheels. They repeat in random order; you can't predict when the next star will appear, but you know it won't be long.

A similar pattern can be found in three dimensions. It had been an article of faith with crystallographers that only certain kinds of symmetry could build a crystal—and five wasn't one of them. Crystals had to be built with three-dimensional shapes that could fit together without any gaps, so they needed to have two-, three-, four- or six-fold symmetry. However, in the past few decades, we have become aware that there is a different kind of material that falls between the amorphous fluidity of glass and the absolute translational symmetry of salt. Quasicrystals are formed of at least two shapes that fit together to fill space completely, without gaps. When the crystallographers make their X-ray diffraction patterns of quasicrystals, the resulting symmetries are often the forbidden number, five.

It makes me think of time building its seamless continuum, dealing us patterns similar to the past but never repeated exactly the same way. As the quote attributed to Mark Twain goes, "history does not repeat itself—but it does rhyme."

FIVE is also an interesting number in poetry—English poetry, anyway.

"Of writing many books there is no end," wrote Elizabeth Barrett Browning at the beginning of *Aurora Leigh*, a line that is a perfect iambic pentameter with its five alternating stresses. Then she swings into thousands of lines, the same recurring pattern tiling page after white page through the nine books of her heroine's story about becoming a poet. Iambic pentameter has been the measure of poets from Chaucer through Milton to Tennyson:

It little profits that an idle king,
By this still hearth, among these barren crags...[30]

"To break the pentameter, that was the first heave," declared Ezra Pound. His modernist project included getting away from this monotonous plod to a way of writing poetry that sounded fresher, more like natural speech. But the pentameter breaks up all the time; it's surprisingly hard to hold it together. It doesn't form little crystalline subunits the way a line with four stresses breaks at the middle into two equal lengths. Five is a prime number and has no divisors except itself and one. Most often, a pentameter line will break into two pieces, one with two feet and the other with three:

By THIS still HEARTH // aMONG these BARren CRAGS.

More rarely, there will be a one-four split; in other words, a kind of rest after the first iambic foot or the fourth. And sometimes you can take the whole line at a five-unit gallop with no pauses. Meanwhile, the individual feet are also breaking up. Stress patterns reverse—it's common to have the first pair of syllables in the line reverse the ta-TUM, and there are always little unstressed syllables tripping in. In fact in the first 165 lines of Milton's *Paradise Lost*, there are actually only two perfect iambic lines.[31]

Yet the pattern makes a grouping we can hold onto easily in our working memory. We handle its ideal symmetry and the tiny asymmetries that deviate from it in our minds as we hold and assess the pattern of a human face.

IF ANYTHING could convince you that there is something "out there," it is symmetry as physicists understand it. Not "out there" in the form of space aliens, but outside our own brains, even beyond our directly experienced world. Symmetry comes as close to Plato's ideal world of unchanging forms as you can get.

Symmetry means something has been saved, conserved, held still. For every symmetry in the laws of physics, there is a corresponding conservation law—some quantity (such as the total energy of a system, or its total electric charge) that must stay the same while the system evolves through time.[32] Such symmetries require the universe to treat all observers equally so that the laws of physics are the same from all vantage points. Early in the twentieth century, scientists realized that the conservation of energy was simply another way of saying that the laws of physics were symmetrical with respect to the axis of time. "We change, the world changes, but the laws of physics remain the same forever. A few mathematical calculations, and conservation of energy trivially follows."[33]

Each of the four great forces has a symmetry associated with it. Gravity, for instance, enforces the equal validity of all observational points anywhere in the universe. The feeling of gravity holding us to Earth and of acceleration dragging us through space are two sides of the same coin. If you are moving, your measurements of the world can be correlated with those of another observer who is standing still. Space and time are connected in a deep symmetry.

Since Einstein, we can no longer think of space and time as a vast unchanging theatre on which we act out our lives. Yet the symmetrical relationships—that everything adds up to a constant—are still there in the background, a framework for the continual translations and reflections and rotations of our lives. We fracture it, we pull patterns out of its magic hat, we create and manipulate, but it remains unbroken.

six Poetry and Scale

Flying changes the scale of things. From up here at 40,000 feet, the immense boreal forest of northern Canada looks like lichen on rock, as though the huge landscape below has been reduced to a boulder. An hour or two from now, when we begin the descent to my northern city, the quarter-sections of cultivated land will become a quilt of human textures, the grain of corduroy or twill. The cars sliding along the highway will become insects and then toys. Then the wheels will bump down, everything will become its "real" size once more, and I'll be home.

Whether child or adult, we are fascinated by changes in scale. Look how long people will spend making model ships—the exquisite tangle of thread that's exactly the right thickness to represent rope—or the bridges for a model railroad. Look at our fascination with bonsai, a whole landscape on a ceramic dish.

When I was a kid waiting for an Ontario spring to arrive, I used to loiter by melting piles of snow at the side of the road, absorbed in the way they created miniature systems of river and waterfall, channel and dam. Years afterward, on my first trip to Jasper in the Rocky Mountains, I had the strangest sensation that my childhood landscape had been blown up to a gigantic scale. The wide valley where the Athabasca River makes its way through braided channels and white cascades tumble down mountainsides was the grown-up version of my old walk to school.

SCALE INVARIANCE, the tendency for certain patterns to look the same as you get closer up or farther away, is a common feature of the natural world: branches on a tree, cloud shapes, coastlines, the hierarchies of bronchioles and alveoli in the lungs. It is also a feature of intriguing mathematical objects known as fractals, which have become increasingly relevant to studying the world's varying phenomena since mathematician Benoit Mandelbrot began working with them in the 1970s.

We are used to the idea of dimensions and the objects that go with them—one-dimensional lines, two-dimensional squares or triangles, three-dimensional cubes, pyramids, spheres. We learn to manipulate such objects mathematically: area equals length times width; radius and circumference can be used to establish the volume of a sphere. We use these

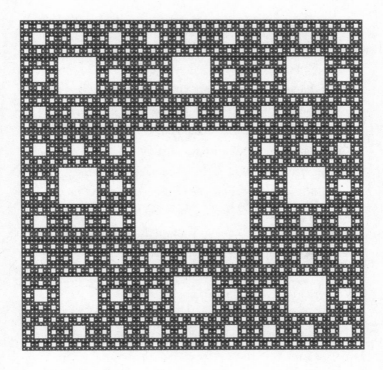

formulae to think about our world and forget that very few
objects are this simple.

"A cloud is not a sphere," wrote Mandelbrot. "A mountain is
not a cone."[1] Lichen covers the rock's two-dimensional surface,
but not completely. The network of blood vessels fills three-
dimensional space but not with solid cubes. Such everyday
examples can be described more accurately using *fractal*
dimensions that are not one, nor two, nor three. Instead, they
fall somewhere in between.

As an example, take the Sierpiński carpet.[2] It's a rather
simple construction built by taking a square and dividing it

into nine smaller squares. Knock out the centre square, then take each of the smaller squares around it and repeat the process. You can continue this dividing process infinitely or you can change directions. Take your original square as part of a larger one, and let the pattern grow into an infinitely large doily knitted by a mathematician god.

The square we started with was a standard two-dimensional shape. However, the area of the Sierpiński carpet does not increase according to the simple formula of length times width. Instead, its area increases at a somewhat lesser rate as the carpet gets bigger. Although it is a two-dimensional object from a topological point of view, its fractal dimension doesn't quite equal two. Instead, it actually has a D (as mathematicians label these in-between dimensions) of 1.89.

Fractals had been noticed as early as 1875, but they were regarded as mathematical monsters that could be safely disregarded as irrelevant to a description of the natural world. Scientists had spent centuries in studying the properties of regular shapes like circles and waves. This colossal investment of effort formed the very foundation of science, and it was felt that minor squiggles could be ignored.

However, while regular curves like the circle are interesting, they are rare and their behaviour quite special. "In comparison, 'wiggles' have been left almost totally untouched," wrote Mandelbrot.[3] He began to find in his research that one of these monsters after another could become a tool with which to answer some old questions we have been asking about the shape of our world.[4]

A fractal dimension is really a measure of the wrinkliness of an object, or of how fast its length, surface or volume will increase if we measure it at ever-decreasing scales. Take a coastline, for instance. We say the coastline of Canada (the enormous country I am flying over) is 243,042 kilometres long.[5] That sounds reassuringly exact. But it's based on a measurement taken at a specific height; the official length in the *Atlas of Canada* is based on a scale of 1:250,000. However, if we moved in closer, the length would increase as we drew our border around increasingly small bays and inlets, a self-similar pattern that could eventually be drawn around every rock and grain of sand. The ideal of a one-dimensional coast "line" twists into a fractal dimension that is typically about 1.25.[6]

Self-similar fractal patterns are ubiquitous and tie together phenomena ranging from metabolic processes inside cells to the phase transition when water turns to ice. They are a way that natural processes can make room for themselves—the fractal nature of coral reefs and tree canopies makes more space available for biological activity and colonization than would be available in the expansion of ordinary dimensions.[7] Fractals tend to occur when new order is coming into being and show up not only in physical systems but also in cultural networks from to the stock exchange to Facebook.[8]

Even our brains are fractal objects. The neocortex is actually a two-dimensional sheet that would be about one metre square if we stretched it out on some imaginary ironing board. To fit inside a skull, its area has to be crumpled into a series of smaller and larger folds until its dimensionality approaches that of a 3-D object.[9]

BECAUSE WE ARE CREATURES who evolved in the natural world, scaling relationships give us intense pleasure. They allow us to recognize a deep interrelatedness in the world's disparate phenomena, where its parts are related both to larger structures and smaller ones. We are amazingly good at extracting fractal relationships from the incoming stream of wavelengths—sound and light—that pour in on us. In fact, recognizing fractal relationships is quite possibly central to how our visual and hearing systems distinguish random noise from meaningful data.

Aesthetically, we are particularly fond of fractals that fall in a certain range of self-similarity, more at the middle. Patterns that repeat too exactly are monotonous; others that don't repeat at all are uninteresting for a different reason. A pure monotone is the auditory equivalent of a straight line—a unidimensional repetition of a single wavelength that could quickly drive us mad. "White" noise, its opposite, is made up of all possible wavelengths. It's the hiss from speakers or an old television, a featureless blanket that we quickly tune out.

Instead, we like the sound of water, where we hear self-similar clusters of wavelengths repeated for shorter and longer periods, or the rustle of leaves in the wind. Visually, our preferences also lie in the mid-range of fractal possibilities—most frequently we like a D of between 1.3 and 1.5.[10] We like the patterning of mountain peaks with smaller and larger versions of the same shapes laid over each other. Or clouds. Or the shifting shadow of leaves on the floor.

"Glory be to God for dappled things," wrote poet-philosopher Gerard Manley Hopkins. His line wouldn't have had read nearly so well if he'd written, "Glory be to God for fractal objects"; but it comes to nearly the same thing.

WE TRANSFER our fondness for fractal relationships to art. Like natural systems, the human constructions that satisfy us do not merely repeat the same elements but also express the same patterns at different scales.

"In all branches of cognitive endeavour, our highest praise is reserved for works that build the deepest hierarchies," writes musicologist Robert Jourdain.[11] Here he is explaining how rhythm helps build musical phrasing, which in turn builds larger compositions. But it applies to the visual arts, as well— think of Hokusai's famous print, *The Great Wave off Kanagawa*, where a huge swirl of water in the foreground curves forward over the distant peak of Mount Fuji, and the cresting wave's shape is echoed and re-echoed in its breaking foam.

Even in art that doesn't represent the fractal shapes of the natural world, our preference shows up. The famous drip paintings of Jackson Pollock were built up of layers of paint; when physicist Richard Taylor analyzed them by breaking the images into smaller and smaller blocks, he found their patterns repeating at all levels.

"[Pollock] described Nature directly. Rather than mimicking Nature, he adopted it is language—fractals—to build his own patterns."[12]

AND POETRY? Any genre that ranges from haiku to epic can obviously exist on many scales. An individual poem is like a coastline—its length increases as you get up close to it. Take for an example Tennyson's "Ulysses," another of the poems I loved when I was a little girl:

> *It little profits that an idle king,*
> *By this still hearth, among these barren crags,*
> *Match'd with an aged wife, I mete and dole*
> *Unequal laws unto a savage race,*
> *That hoard, and sleep, and feed, and know not me.*

In those first few lines, you are flying just high enough to see the surging pattern of iambic waves, their spacing as they roll to shore. Fly a little lower and the details of the individual line break into a foam of sound.

Within that foam, we notice certain repeated shapes. Humans have a lot of unconscious expertise in analyzing the frequency of sounds in language, and we know when we're hearing a pattern that is slightly atypical—a promontory or prominent cliff. For instance, the consonants made with the tip of the tongue (like "t," "d," "s," "l," "r" and "n") are the most frequently occurring in English.[13] In the first line of "Ulysses," those tongue-tip sounds predominate until the end, when the consonants of "king" shift into the back of the throat.

Then there are the vowels, the bays and inlets of voice. The most common vowel sound in English is that "uh" sound you hear in "the," which linguists call a "schwa." It's just a huff of voiced air to separate syllables and hardly deserves to be called a vowel at all. It appears in almost all unstressed syllables. As readers, we can be fooled into thinking there are lots of "a" and "o" and "u" sounds in our words. But, while the letter "o" may occur twice in the word "sonority," the vowel sound "o" only comes once: suh-NOR-uh-tee. (But hear how the "aw" sound shifts place when you shift the stress pattern: "sonorous" becomes SON-uh-RUS).

All vowels sound in our ears as a series of formants or overtones, but front vowels (like the "i" of "bit" or "ee" of "seen") are coded acoustically in a different way from the low back ones like "ah" or "aw."[14] Their overtones are more widely separated than they are in the back vowels, where the formants lie much closer together on a spectrogram, so high front vowels tend to stick out in a different way.

Let's go back to that first line of "Ulysses" and take note of the short "i" sound recurring at the beginning of the line: "It little profits" and then again ending the line in "king." At the beginning, the "it" combination combines a relatively noticeable high front vowel with one of the most common consonants in our language. But in the last word of the line, the "i" is paired with a relatively infrequent consonant, "ng," which needs the back of the throat to resonate. A shift has happened, from the little and local to a more extensive exotic sound.

This is the kind of sonic coastline that we navigate in hearing poetry. But like most cartographers, we find this level of detail to be more than we need for a map-making expedition. So we lift back up to the scale where wavelets and waves take on a pattern like grained leather. Here you can begin to see the coastline of story, the curved arc of narrative—Ulysses stuck on the island that he had spent ten long years trying to get home to. And then the plane is lifting higher and you are looking down at a topography of emotion—the human urge to push away from shore, the ache of being simultaneously at home and a foreigner. The poem is linked on all these scales—sound, story, heart.

I WAS FLYING HOME from a conference of the League of Canadian Poets in Toronto. It was early summer 2005, and I had a snug secret in my pocket. I had taken a phone call the previous day from the mayor of Edmonton, confirming that I had been selected as the city's first poet laureate.

My emotions were mixed. When the idea of establishing the post had first been mentioned casually, I felt cool to the idea. Of course it is important to pay attention to an art form that typically ekes out a lichen-like existence on inhospitable tundra. But the notion of banging one poet on the head with a civic club and saying "You're it" went against the grain of what poetry is to me. It seemed to play into the whole cult of celebrity, making poetry a kind of reality-TV game show. Who will struggle through when everyone else is thrown off the island?

But then, when the local political will to establish a poet laureate coalesced into reality, my ideals deserted me. I was urged by friends to put my name forward as a candidate, and when I did so, I found that I *wanted* that honour.

I was to be interviewed by a jury for the post; they would quiz me on my ideas about the position. Please be prepared to read a poem aloud for the jurors, I was told. But because I was away for the conference on the necessary date, I had to phone in at a designated time.

The interview experience fell into a fractal dimension somewhere between epic and comic verse. In an obsessively interconnected culture, making a long-distance call may not seem like a big deal. But poets do not meet in posh conference centres with videoconferencing technology. I was staying in a phoneless dorm room in Victoria College's Burwash Hall. The League meeting was in the stony-pillared vaults of Hart House.

I was sternly refused permission to make a long-distance call from any of its offices. I did not own a cell phone.

I didn't exactly want to make the call from a public phone booth, so I jumped in a cab, having borrowed the key to the League's office—only to run into the snarl of Toronto's downtown rush hour. The cab lurched into a line of traffic and sat there clucking like a constipated hen. After ten sweating minutes, I flung a bill at the driver, leapt out and galloped back to Victoria College on foot. The offices with their forbidden phones were now all locked up for the business day. I pleaded for someone, anyone, to point me to a place where I could make a relatively private phone call. For god's sake, I was looking for a *phone* not a personal teleportation device.

They directed me back to the basement of Burwash Hall, where a phone was mounted on the wall in a bare booming corridor just outside the dorm's laundry room. The long white walls made an appropriate setting for a Hitchcock film—god only knows what might be tumbling behind the glassy occularity of the dryer doors. It felt as though I should be dialing 911 and hoarsely mouthing "help, help" into the mouthpiece instead of conducting a long-distance poetry reading. My voice came out in an un-laureate-like croak when I finally connected and introduced myself to the receptionist, who sounded dubious about accepting reverse charges from a frog in an echo chamber.

"Please hold," she said. "The jury isn't quite ready for you yet."

I waited the long minutes on the phone, reading over the lines of a poem to get used to the sound of my voice in this theatre of the absurd and praying that no one would come out of the laundry room with a basket of sheets or a revolver while I was declaiming. The things I do for art.

But perhaps the weird acoustics stood me in good stead. For here I was, flying back home as the soon-to-be-official-civic-poet. The plane may have been at 30,000 feet, but I was probably 5,000 feet above that. Yet my emotions were *still* mixed. Yes, I'd wanted it. But part of me could also imagine the yuck-yuck reaction from the world's bigger ponds: "Edmonton? Where the hell's that? And they think they have *poets* there?" In vain, I tell myself that my city's population is about three times the size of Shakespeare's London and we've got as much right to poets as anywhere. But the insecurity lingers, and the problem is a fractal one of scale.

YEARS EARLIER, I'd come to this city that would become home from the other direction—from a much smaller city even farther west, where I'd worked as a newspaper reporter for a couple of years. In Williams Lake, British Columbia, I had covered city council, sitting in on evening meetings that staggered on until after midnight. The drama was frequently high when the assortment of elected personalities debated such items as whether or not to replace the official float used in the annual Stampede Parade—a giant, inflatable bull, a goofy bovine that billowed along Borland Street and was prone to slowly buckling as air leaked from its aging hide.

"Council cowed, bull bounced" was the headline I gleefully wrote for that story. It made the front page.

However, my very favourite story as a reporter concerned the beaver dam debacle. Williams Lake—the lake itself—was dammed at its exit by enterprising rodents. As spring began melting snow from the surrounding hills every year, the lake levels backed up and up, getting ominously close to the front doors of houses built along its shore. The province's fish-and-wildlife protectors did not want to do anything to destroy the dam because eventually the cold meltwater would descend to the bottom of the lake and flush up the warmer, stagnant layers below. This stale water would then pour over the top of the dam as part of a natural cycle of renewal.

The hell with flushing out lake water, says Mayor Tom—that dam will flush out people's basements. So he goes out himself in a canoe one night with a couple of sticks of dynamite to blow a hole in a dam.

Oh, we had even more fun writing the headlines that week.

I felt that this kind of escapade must be typical of small-town politics. Surely the assemblies of larger communities must act with more sense of proportion, more sophisticated analysis of issues. I came to Edmonton thinking that a city of three-quarters of a million people would be governed by a city council somewhat more dignified in its proceedings. However, I found the human dimension doesn't change much. There were huge uproars when a new mayor didn't want to wear a large beaver hide that came with the official chain of office. For her, it was a statement about animal rights; for the rest of council it was a rejection of the city's history as a great fur-trading post.

For me, it was more of a fashion statement—wearing that beaver hide is like having a great round furry pond draped over your shoulders from which your head sticks up like a lonely lotus bud. Only a mayor built like a buffalo can carry it off.

The whole silly debate occupied more column inches in the paper than the approval of millions of dollars in road repairs. You might think this sort of thing happens just because we're out here on the lone prai-ree and the winters are long. But then you watch televised proceedings from the national capital of the federal government's question period, in which the level of discussion is hardly higher. Human beings don't change scale much, regardless of the size of the stage they walk on. The distribution of capacities and talents is much the same in any group of humans. Making the group larger doesn't lead to a corresponding increase in the individual IQs clustered under the bell curve. To this extent, human beings are *not* fractal.

HOWEVER, assemble us into progressively larger groups and fractal patterns do tend to emerge. The qualities that make a *good* poet are a complex of linguistic ability, creativity and the desire to invest the time and effort necessary to succeed; these qualities are more or less uniformly scattered through the population. The qualities that make a *famous* poet are not so different, but they are compounded with something that can only be described as luck. The process of poetic fame is governed by fractal patterns.

Nassim Nicholas Taleb, in his essay "The Roots of Unfairness: The Black Swan in Arts and Literature," points out that literary or academic fame is analogous to stock-market booms. There's a winner-take-all effect. In any given year, one book in eight hundred will account for half of sales, while the other 799 eke out a meagre share of the pickings. We jump hopefully into the publishing cab and go nowhere much.[15]

The phenomenon operates much like the Mississippi River system scooping up the water from smaller and larger tributaries in an immense drainage basin. It becomes impossible for a water droplet to cut its own channel to the sea. The pattern is not caused by malice or design or even by commercial greed. Taleb points out that the same pattern emerges with the academic citation system, supposedly free from commercial interests. If you are the lucky researcher whose paper first gets cited out of all the researchers who may be working on the same problem, you'll go on getting cited by all the future researchers. And the bigger the drainage basin, (i.e., the larger the number of contributors like authors or researchers) the higher the concentration into one main channel will be.

An idealized mathematical process like the Sierpiński carpet can be subdivided forever. However, the real world is not scale-independent in that way. In our world, most things are not subject to one fractal pattern but to two or more simultaneously. Such multifractals tend to come to a natural limit.[16] Patterns that work at small sizes don't work at large ones; hierarchies emerge in response to physical constraints. The early days of an embryo's existence can be nourished without a central circulatory system, but fairly soon that doesn't work any longer and cells need to specialize.

In poetry, the physical constraint we are up against is time—the public's time. In any one life, there's only so much time to read books. Few citizens can take on three hundred poetry books, the number published annually in Canada alone. You need a process for deciding which ones are worth your while. In isolated tribal systems, there are only so many bards, so many works, and each of them can be absorbed and recognized. In large urban societies, we depend on some kind of filtering system.

Fame is just what happens when you can't know every book personally—when you have to fly at 40,000 feet to get across the country in a manageable amount of time. Its filtration system is essentially an information-exchange process made up of little magazines, poetry contests and prizes, and includes the luck of proximity to the whole tag team of mentors, publishers and reviewers. It's less hierarchy than swamp bed or compost pile. Only those poems or poets who are very sturdy, accidentally lucky or both will survive.

Of course, it's ironic to want fame in poetry. If you want to be famous, go into rock music or write self-help books. Only a tiny number of Canadian citizens can name a Canadian poet, famous or not. Still, it's only human to want to exist on as large a scale as possible. The unkind cut about being satisfied as a big fish in a small pond overlooks our need to live in a pond that's small enough for us to be noticed, respected, held in a collective consciousness. It's debilitating and discouraging to feel that all the attention is being sucked up by a few authors who aren't even from around here—that the book readers and buyers in your own city will be more familiar with the name of a highly promoted author from the east (whose book isn't so much better than yours that it deserves *all* those reviews.)

The great advantage of being a poet is that the filtration system is so damn slow. A novelist will usually have a one-time chance to cut a channel with her book. For the poet, reputation is more marsh than Mississippi—an ecology that lets a lot of us flourish locally, which is really where we're needed anyway. And poets can always hope to be dredged up after they're dead. It's one of the small compensations of our vocational choice. What actor can hope to finally win that terrific role after he dies? What athlete will rise to the Olympic podium from six feet under?

SO, restless ambitious young poets often give up on poetry to work on fiction. They can hardly help it. "When are you going to write a novel?" they are asked by well-meaning friends who want them to succeed. Poetry is thought of as a practice run. It's short—surely you intend to write something bigger, more saleable.

Instead, I've closed down. My first book *was* a novel, and I expected I would return to fiction. In the difficult years of finding a publisher for my first book of poetry, I wrote sadly in my journal that I'd probably always write poetry "on the side." But the collection did come out (I had to help start a publishing company to make that happen) and slowly I found that this is what I wanted to bind myself to, after the flirtatious early years.

For me, poetry is a bit like the deep-field pictures of the universe collected through the Hubble Space Telescope. Through a tiny aperture, the size of a grain of sand held at arm's length, you can see deep, deep, deep into things, into how we came to be, farther into space and time than is possible in more comprehensive surveys. And you bring back photographs for others.

WHICH BRINGS ME TO cosmology, because the big headache of the last century has been to link the very large with the very small. To study great big things—planetary orbits, galaxies, the universe—physicists have used the mathematics of relativity. To study the smaller and smaller scales of atoms and photons and quarks, they've used the statistical patterns of quantum mechanics. The trouble is, the two systems have components that just won't fit with each other. Relativity is based on smoothly curving quantities that vary continuously; quantum mechanics has gaps, things that jump from one state to another with nothing in between.

Yet cementing the snake's head to its tail is essential if we are to understand the ouroboros. Physicists have been trying to do this through approaches called "string theory" or "loop quantum gravity." One of the intriguing things to me is that such approaches throw up the idea of reciprocal dimensions. If one set of mathematical tools make a certain dimension large, its reciprocal partner dimension will be small. If you switch to using a different mathematical approach, those relationships reverse—what was small becomes large and vice versa. Both dimensions are simultaneously large and small; what you see depends on the set of equations you're looking at them with.[17]

Well, poets have known this kind of thing for a long time. In poetry, the very near is the study of the very far. But poetry can't *seem* small. It may be pulled very tight, but it has to pack a lot. And it can't pack just a single dimension, which is why a great many poems can seem inadequate. They develop a kind of quantum claustrophobia when the world is no larger than the poet's kitchen. Poems about having coffee, poems about writing retreats, poems about other poets—we churn them out, but they remain small, limited.

Yet the poet's kitchen can inhabit a dimension that is as large as the universe. In the tiny curled-up dimensions of string theory, to a particle of that dimension, its universe is as large as the more familiar-scaled universe seems to a photon.

ONE OF THE THINGS I often dislike about myself is a constant wondering if I will "win," if I will be "big enough" as a poet to make this whole life's work worthwhile. It's the feeling that somehow there is a path to acceptance that will take away the ache of being an outsider. But we are always outsiders, all of us.

"Inside" is a myth. Even if you feel for a little while that you are inside, that you are an important part of a group, there is always a larger structure in which you are a stranger. Worm your way into that and there's a larger one still, until you are faced with the paradoxical loneliness of God—that which is forever beyond and apart from us, yet that something to which we belong, utterly.

We always exist at the edge, the circumference, a perimeter. It's like the edge of the Mandelbrot set, a region in which self-similar patterns emerge as we zoom in closer and closer to its mathematical country—sprinkled curves and arabesques, small lakes of belonging and the broken coastline of longing. But fortunately, we live in a world where the fractals come to a limit. The plane bumps down and we are home.

SEVEN

The Ultraviolet Catastrophe

I fell in love with the phrase as soon as
I read it: "the ultraviolet catastrophe." All
those repeated "t" and "tr" sounds and the
rhythm and, of course, the connotations of
the words themselves. For years, I've been
trying to find a poem to go with it. About a
thunderstorm? A rock concert? Love?

It would have to be a poem about an event that threatens but never happens—the lightning doesn't strike, the stranger across a crowded room turns out to be your gynecologist. The ultraviolet catastrophe was, for physicists of the nineteenth century, something that, surprisingly, did *not* occur. It became a kind of test case, a small pinhole in their picture of the world that, when looked through, revealed a completely new understanding.

The ultraviolet catastrophe that did not happen concerned some very ordinary behaviour in the world, as ordinary as how a poker glows in the fire. At the end of the nineteenth century, the scientific picture of the world included two forms of existence: very small particles, and fields that moved energy in continuous waves. But the deep mathematics of this picture didn't quite work out. Scientists predicted that, as you applied more and more energy to the particles, this energy should radiate away at continuously higher wavelengths. The radiation from an increasingly heated object would climb through the colour scale from low-energy red wavelengths to the high-frequency violet end of the spectrum and beyond, to the invisible energies of ultraviolet light.[1] But however many pokers sat in fireplaces all over Europe, they never turned violet as you heated them to still higher temperatures. They maintained the same orange or white glow.

This was a clue that light should be thought of not only as a continuous electromagnetic field but *also* as being made up of particles. Einstein made use of a piece of mathematical "trickery" developed by Max Planck to explain light as small lumps of energy—photons—defined by their frequency. An orange photon stays orange—the poker sends out more of them as it gets hotter, but the increased heat energy doesn't alter their frequency. Different substances emit photons with different wavelengths; the characteristic yellow-white light of the sun is indicative of its particular composition. Einstein's paper on the photoelectric effect elegantly explained a nice little puzzle, earned him his Nobel Prize in Physics and opened the first window into the counter-intuitive world of quantum physics.[2]

VIOLET is that shade at the extreme inner edge of the rainbow, the colour that seems more like the rainbow's shadow than a shade. It is made of photons with wavelengths around 380 to 450 nanometres, the shortest wavelengths our eyes can see and the first step in the huge range of ultraviolet radiation that lies beyond our ordinary sight.

But, because my father was a house painter, I thought of violet not as an edge colour but as an inside, secondary one. He taught me the colour wheel when I was little, and how certain colours—red, blue and yellow—were primary, the sources for other colours. Violet (or purple) wasn't the extreme opposite of red at the far edge of the rainbow; it was its close neighbour on the colour wheel. I mixed little tablets of scarlet lake and ultramarine in my paint box together to get purple, and watched the water glass where I dipped my brush turn the dull tertiary colours—greenish brown, puce mud.

I found it confusing in high school when I learned that mixing *light* wasn't the same as mixing pigment. Suddenly yellow became a secondary colour, not a primary. Disks of red and green light burst into yellow where they overlapped. And when you added a third overlapping disc of blue, you got white light instead of the muddy almost-black of my water jar. It took me a long time to understand that the paints and pigments my father stirred together made their different shades by *subtracting* wavelengths from the white light that bounced off a painted surface; eventually mixing enough colours would subtract all the wavelengths and leave you with black. Conversely, mixing together the light shining through different-coloured filters added their wavelengths together, returned the spectrum to white.

What was a primary colour or a secondary one? What was a pure colour or a mixture? What was on the edge of the rainbow or inside the colour wheel? Colour was far less simple than my paint box made it seem.

VIOLET—OR MAUVE as it was fashionably dubbed by the Victorians—was the first of the aniline dyes. Purple is a difficult colour to get from natural sources. In Roman times, it had been manufactured from shellfish of the genus *murex*, but the ancient process for making "Tyrean purple" had faded into the ultraviolet realm of forgetfulness. In 1856 an eighteen-year-old chemistry enthusiast called William Perkins was about to throw away the black residue left from an attempt to make synthetic quinine from coal tar when he noticed the strangely beautiful colour the substance made in solution.[3] Within two years, Perkins was a rich man—mauve was wildly popular, his dye factory was colouring the dresses of Europe and chemists were racing to extract more and more brilliant shades from coal.

Without his discovery, industrial dyers would have continued to create purple from a blend of indigo and madder or from lichens, but nothing delivered the knock-out purple blast of the aniline dye. This is because the synthetic dyes represent a kind of extreme: their brilliance comes from having only one wavelength, only one colour. Natural dyes vibrate in more complex tones. When you look closely at a piece of madder root, it contains not only its distinctive bright pinkish red, but a whole range of other colours.

Life and literature are like the natural dyes—they do not lend themselves to extremes. We cannot live at such unadulterated wavelengths. But there are points where we approach them.

THE UNIVERSE doesn't give us edges any more than the colour wheel does. As the Greek philosopher Archytas pointed out more than 2,000 years ago, there couldn't be a boundary: "If I am at the extremity of the heaven of the fixed stars, can I stretch out my hand or my staff?"[4] He realized that you could never poke a staff through an edge of the universe. There would have to be something beyond it. The mathematics of contemporary physics confirms his intuition: difficult as an infinite universe is to imagine, it is still more difficult to imagine something outside it.

Nevertheless, the universe does give us extremes. A "singularity" occurs when, if you concentrate enough mass in a small enough space, the ordinary relations between matter and energy break down and simplify. The force of gravity always adds up—it is weak compared to the other forces, but it doesn't cancel itself out as electromagnetic particles do, or fade away over very short distances as the strong force does. Eventually it pulls the matter closer and closer into an infinitely small space, a point where gravity itself becomes infinite.

When the implications of this tendency of gravity were noticed by astrophysicists in the 1920s, they were generally considered to be a fiction of mathematics that wouldn't occur in the real world. Sir Arthur Eddington mocked the young Indian scientist, Subrahmanyan Chandrasekhar, for his idea that stars would have to collapse.[5] Something must intervene to keep this from happening. But the idea of black holes (or frozen stars, as they were first called) didn't go away, and the calculations eventually proved Eddington wrong. The normal life cycle of larger stars concludes with them collapsing into singularities. When astronomers began looking, they found that they are a rather ordinary feature of the universe. There seems to be such a black hole at the centre of every galaxy, including our own, pulling matter into its invisible fist. These are the end points of the universe, but they are scattered throughout its interior.

A black hole is a simple object.[6] It can be defined by three parameters—how big it is, what its electric charge is, and its rotation. All the varied characteristics of the matter that is poured into it disappear; the segregation of quarks and electrons, protons and photons that happened in the early days of the universe is reversed, smoothed out. Its gravitational force is so enormous that it even traps light; you can't see a black hole if you are beyond a radius known as its event horizon. All you can witness is a faint fizz of radiation around it and the tendency for its invisible mass to tug and distort matter and energy in its vicinity. At its centre, the mutable multihued world is reduced to colourless unchanging force.

WHAT ARE the edges of tragedy? Where does catastrophe begin and end?

"The comic is often paired with the tragic, but the two concepts are asymmetrical and different in kind," writes Iris Murdoch in her perceptive analysis.[7] We think of the two as extremes of the rainbow spectrum, deep violet mourning opposed to the jubilant red of the clown's nose. But of course, as philosophers and clowns have known throughout history, gladness can lie achingly close to pain on the colour wheel of emotion.

"Tragedy belongs only to art, where it occupies a very small area," writes Murdoch, pointing out that we have been trying to define its essential characteristics for generations. It's big, say some; it needs kings and princes. It's about hubris, a fatal flaw in the main character, say others. It should be banished, wrote Plato in semi-mock sincerity, since it charms us into identifying with bad and extreme characters.

Murdoch questions whether the attempt at isolating the particular colour of tragedy is even a particularly useful effort. "Is there not something wilful in the attempt to define tragedy, making it out to be something interesting and ideal? Why not treat the tragic plays as multifarious works, full of aspects and ideas and multifarious stuff?" she asks. The concept of tragedy "indicates some kind of end-point, a remarkable break in the continuum of art," says Murdoch, pointing out how few works compose this singular genre: some plays from the Greeks, Shakespeare's *King Lear*, a handful of others.

What kind of artistic misery counts as tragic rather than as pathetic or sad? And why on earth would we want to put this kind of pain into art and feel that it provides us with release, with the emotion we call "catharsis"?

Tragedy involves a cluster of features. There's death of course. "That someone must die in a tragedy is not a mere

convention, like that which decrees deaths in detective stories,"
writes Murdoch. Tragedy involves a quality of undeservedness,
unfairness, of innocence in collision with evil. It also involves
a kind of compressive inevitability, a train wreck happening
that can't be stopped. Antigone cannot resolve the conflicting
duties that will lead to her death. There is a reduction in the
number of dimensions of life, with everything sucked into the
crushing gravity of a black hole, down to one or two defining
emotions. There is an unrelieved quality—any consolation at
the end is feeble and second-hand.

But, unlike Iris Murdoch, I don't think tragedy belongs only
to art. It is a central emotion; it wells from the way that human
beings experience grief.

The brain circuits underlying adult grief are based on those
of an infant experiencing separation anxiety. This set of
emotions becomes available to babies from the age of about six
months; by this time they will likely have developed a bond
with a mother/caregiver that is unique and close to irreplace-
able.[8] In all mammals, separation results in a well-identified
series of behaviours that begins with an initial period of
distress and repetitive searching: an infant will become actively
angry and hostile toward her mother for leaving and at anyone
else trying to substitute for her. In this immediate phase of
searching, small monkeys experience elevated heart rates,
increased body temperatures and sleep disturbances. This first
flush of protest is succeeded by an intense phase of grief, in
which body temperature and heart rates drop; little monkeys
will take on the curled-up body positions and sad facial
expressions that humans do. Hormone levels change.

After these initial reactions comes the longer phase of
mourning and adjustment. "Grieving" we call it, this long

period of lowered mood, disturbed sleep and cognitive circling around a dark centre. This is what our self-help books and psychiatric interventions focus on, how to get us through/over/around this period. But this period is not what tragedy evokes. Instead, it's about our initial desperate capacity for distress, a phase of such intensity that we are afraid our capacity for pain could kill us, too.

We carry the infant's fear of abandonment forward into the adult understanding of death—a knowledge that grows only slowly throughout childhood. Adults are able to fear death in advance, come to understand its extreme finality. And this is the emotion that lies at the heart of tragic literature. Tragedy is not about kings and princes per se; while they are often the subject of the classic dramas, they are a metaphor for the importance of the people we will lose.

The classic tragedies involve a combination of innocence—someone is hurt who doesn't have to be and didn't deserve it—and inevitability. We cannot help dying, we cannot help being left behind. If someone has actively caused that separation—even the missing parent—he or she is a suitable object for hostility. Wilful evil has a role in tragedy but not a more central one than it plays in fairy tales. It's a plot device as much as anything, something to get the pain going. Hubris, or other flaws that trip a character up, are also as characteristic of comedy as of tragedy. What's different in tragedy is the context of unrelieved pain and the question of whether it can be survived. The experience of catharsis is not so much the purging of emotions as if they were some kind of boil, but the reassurance that we do survive this singularity, this near-approach to almost-infinite pain. We may be left with the ashen world of depression, kneeling on a stage with the body of someone we love in our arms, but we have gone to that limit and lived.

MEMENTO MORI and *carpe diem* are two of the great themes of poetry, flip sides of the same coin. Remember death and seize the day. These are the frames we put around experience, the knowledge of the vibrant spectral band of visible light and the darkness that edges it in either direction.

All poets come to elegy at last, whether it's the psalmist who reflects: for humanity, our days are as grass, we flourish and wither as a flower of the field; or a rap artist who mourns a murdered fellow rapper. However, elegy is not tragedy, not written from the centre of catastrophe. That's a place art can seldom get to. Elegy comes from the complex inside of the colour wheel where what was primary, at the far edge of our capacity to feel, becomes secondary, a mixture.

I WAS SITTING at the far end of an amalgamation—two long tables shoved together for a dinner to wrap up a symposium on the arts and sciences. One of the symposia guests was Alan Lightman, the physicist-turned-novelist, who had given an inspirational talk to open the conference the day before. A cluster of people from the University of Alberta's physics department surrounded him on both sides of the table, and I felt I was watching a lighted stage filled with larger-than-life characters. One of the distinguished older physicists, whose name slid from my Teflon brain as soon as I was handed it in the introductions, had a big energetic beard like a reverse whitecap curling forward from a wave. Largely, he ordered the appetizer *and* the dessert, and the table's third bottle of wine, then settled down to reminisce about the days when he had been working with Lightman, down the hall from Richard Feynman. They told the story of Feynman drawing equations on a board in one of their offices to demonstrate the evaporation of a static black hole. This was before Stephen Hawking picked up the idea and elaborated on it to include rotating black holes, a discovery that made Hawking's reputation. But Richard Feynman had merely dusted chalk from his fingertips and walked away from his equations.

It was like listening to legends of the gods. I was awestruck at being taken to within one or two degrees of separation from Hawking and Feynman. And yet it made me realize how ordinary and human it all is. The world of international theoretical physics isn't much bigger than, say, the world of North American poets. Its members can know each other to within one or two degrees of separation. Their work is a constant dialogue of influence and disagreement. They create their heroes in a process that is partly the luck of history's narrative and partly based on fact—Mount Everest really is a little higher than all the other mountains around it, even if it was created through the same processes and couldn't exist without them.

So as a little foothill at the far end of the table, a ripple in the poetry range, I looked along to the middle peaks of another cordillera and felt how we all stand on the same ground.

(Can foothills have feet? How fractal a thought...)

STEPHEN HAWKING is an extreme, one of the most recognizable physicists of our time or any other. We know him not just for his ideas but for his story—the image of his wasted face and body and the sound of his synthesized voice seem to push us to the edge of human capabilities. How can a body be capable of so little and yet a mind be capable of so much?

He reminds me of my niece, Monica, who was diagnosed with a particular form of muscular dystrophy—fascio-scapular-humeral—when she was a little girl. The condition prevents her body from making muscle tissue. With so little muscle, her bones must carry all the structural weight of her body, which has pulled her spine into a curve. From being a sturdy adventurous toddler, she became a young woman confined to a special wheelchair. She cannot sit up and so lies on it, rather as if it were a high-tech, adjustable surfboard. She spends her life in this position, yet she has graduated from art school. With her impossibly slender fingers, she creates funny evocative drawings, including a cartoon character called Silly, her alter ego. Monica's case is exceptionally rare. This particular form of muscular dystrophy usually begins much later in life and develops to a much lesser degree. She is evidence that the urge to make art cannot be confined, even though it is created on a smaller physical scale.

Our world is dotted with such singularities, individuals like Hawking or Monica or the neurological extremes described by cognitive scientists like V.S. Ramachandran and Oliver Sacks. Sacks describes the case of Knut Nordby, a Norwegian scientist with achromatopsia—a complete lack of the cone cells that give us colour vision—who must depend on the light-sensitive rod cells that help us to see in low-light times of evening and night.[9] He is not simply colour blind. He cannot bear bright daylight without dark glasses to shield him and needs powerful magnification to enable him to read. And yet, though a star should be too small a point for him to make out, he can easily see the constellations of the night sky. And he does not experience his world as colourless and incomplete. He enjoys the visual world and experiences beauty as much as anyone does.

How often our stories are about extreme situations. Oh, yes, there is a genre of literature about the ordinary, the tale of someone who gets up and has eggs for breakfast and walks the dog and collects the mail. But the stories we love are of the un-ordinary—the crisis, the faraway place, the exotic character. I think that's because these extremes are rather like the extreme cases of neuronal malfunction described by Oliver Sacks; in learning about the man who mistakes his wife for a hat, we glimpse how the brain works normally.

Humans are fascinated by extremes. This is the material for our stories, the stuff of our legends. We don't really find the ordinary terribly exciting. We seem to find that such singularities illuminate the human condition. The outliers at the far ends of the bell curve are diagnostic in a way that the bulk clustered under the centre, the standard deviations, are not. They illuminate our understanding of the more ordinary.

People like Hawking and Monica are extreme cases of the basic human condition: the body both limits and enables us. We are all subject to death; we can none of us work beyond the power of our muscles. But when I see Monica's index finger, long and elegant, hover over the keyboard as if the muscles were being consciously manipulated by a puppet master, it brings this central human fact home. And then that her mind is so completely unaffected by the disease, that it can think as deeply and creatively as anyone else on the planet—well, it almost makes you think of mind-body duality. The mind is so completely dependent on the physical to exist and yet somehow deeply independent of it, as well. So people like Stephen Hawking tell us a lot about what it is to be human. We ponder what it means to have so much taken away from our lives. What is essential? What is not?

"Isn't that tragic," we sometimes say of people like Monica and Hawking, which irritates Monica intensely. She would rather not have been dumped with this particular set of problems, but she does not see herself or her life as a tragedy. She gets up every morning. "Being in a wheelchair couldn't stop me, not even the odd flat tire. It certainly couldn't stop me owning a killer wardrobe either," she writes.

TWO SHORT MONTHS before that dinner with the physicists, I had watched my father die. He had become ill so suddenly, a sudden flood of septicemia into his bloodstream from a relatively routine infection. No one else could get here in time. It was just the two of us in a tired room in a dementia ward. The hot dawn sky had turned a strange colour—a roasted bronze with a copper haze. I wonder how he would have mixed that exact shade, my father who could match any colour for the most demanding customer.

Thirty-six hours before, I had made a regular visit. He had been a bit more wobbly, more unco-ordinated than usual, but I wasn't unduly worried. He had one of the gastrointestinal viruses that went round and round the ward like a DNA chain. I had played a CD for him of Robert Burns's songs—ending with that great poem to human equality, "Is There for Honest Poverty." "For a' that, an' a' that, a man's a man, for a' that." As

I turned the music off and bent over him to say goodbye, he murmured, "Bring me a pen. Tomorrow."

I came back the next day to find him lying as if asleep, breathing like a man in a long race, fluid melting out of his cells. Suddenly, as though a sentence had been completed in our heads, we recognized that he was very, very ill. The long summer evening and its short night wore away; the sun rose at five in the morning as I sat beside him. An hour or two later, I came back in the room after waiting outside while the nurses changed his sweat-drenched linen yet again. His breathing was still rapid but less noisy. His face had changed. I took his hand and talked. A nurse came in and said, "Oh, my, he's going," and left us again.

I kept talking. I told him how much he had been loved. I told him the stories of poverty and feistiness. I told him about going up into the hills with his father, to dip a cup of water from the Pappert Well above Balloch and that he'd be going back there to find the three trees young and shining. I sang him the old songs—"Dream Angus" and "By Yon Bonnie Banks"— and was just barely able to choke out "Mairie's Wedding," the song sung at his own wedding. I told him about how he'd met Wee Mary Matheson, and all over again how much he was loved. The breathing was becoming still quieter, but I kept my eyes fixed on his. It seemed as though he was looking at me, though that might have been a wished illusion, a trick of light. I told him about his honeymoon on the Isle of Luing, and then I recited part of one of his own poems, written to my mother:

and we laughed and lay in the heather,
and how I loved you there.

And just at that moment, I realized the breath had stopped. The lid of his left eye slid closed. I watched a moment or two longer to be sure, then I stroked the other eyelid shut, as he had done for his own father more than fifty years before, and sat holding his hand.

"CATASTROPHE THEORY" is the name for a kind of mathematical analysis developed in the 1970s to describe the ways in which a system crosses from one stable equilibrium state to another.[10] It applies to dissipative systems—a pendulum, a steam engine, a human body—that exchange energy with a larger environment. These are systems that will eventually run down, reach some rest condition or equilibrium.

Such a system can have more than one equilibrium state. Imagine a lake in a mountain landscape. The water in it may be held by a scoop of geology in a basin high above the river far below. It is at rest there. But if the lip of rock that holds it in place wears away, molecule by molecule, the system will reach a cusp, a state of instability. And suddenly the boundary may be breached; the water will rush down to the lower basin. A major change to the system has resulted from one last, minute change in the external variables that affect it.

Catastrophe theory relies on deep mathematical models related to how singularities are classified. But the theory does not enable us to make precise quantitative predictions of the future. It cannot tell us exactly what route a water droplet poised at the top of a divide will take, which basin of attraction it will fall into among the available options. It only tells us that there are patterns of change and we can recognize them after the fact.

LIFE IS A BASIN OF ATTRACTION, a tendency toward a stable self-perpetuating equilibrium, a lake cupped for a while in a mountainous geography. In this sense, my father's death was catastrophic, a draining away, a sudden change to an alternative state of equilibrium. Was it tragic?

In some senses, yes. Alzheimer's had been a cruel joke on a creative man. After all the years of struggling to raise a family, he had begun to write more seriously; he had taken some classes, written a wonderful brief memoir about his father and had started to work on a novel based on his experiences as a gunner in the Merchant Marine during the Second World War. He had the skills and storytelling ability to make it work. But then, about fifty pages in, he suddenly wasn't e-mailing us new episodes anymore. He seemed curiously stuck. It was a while before I realized it wasn't just writer's block. He couldn't hold the story in his head to work on it. We didn't understand this was the first early warning sign of more than a decade of pervasive cognitive loss. Had he been given only a couple of years longer before the disease set in, he would have managed that book—I know he would—and we would have it now, rescued from forgetfulness. And in some ways, you might say he was tragically flawed. He *could* have spent his earlier years in a more focused way, with less alcohol, creating more, sooner.

In some ways, no. His life wasn't about his success as a poet, and those were not his only gifts. And my father's quiet death was a release. That last walk behind his body, past the blank faces and clutching hands of the dementia ward, filled me with relief. Never again would I have to leave him there.

Grief, the sensation of tragic pain, was a delayed wave that only hit several days later. The ocean floor has shifted, far out to sea, and the flooding impact finally arrived. I huddled like a small monkey in a cruel experiment, weeping for the young father I suddenly remembered and needed, who had abandoned me.

WE SHOULD SEE the sky as violet, not as blue. The sky's colour depends on the fact that the atmosphere scatters photons differently, depending on their wavelengths; photons with shorter wavelengths like blue or violet bounce around more than those with longer wavelengths like orange or red. When spectrometers analyze skylight, they find it has the greatest intensity in the violet wavelength. But human eyes, unlike spectrometers, do not register single wavelengths.[11] Our brains combine the effect of all the photons that register on our eyes, and we are not very sensitive to violet compared with other frequencies. So we register the dominant hue of the sky as blue.

Something has to be very, very uniformly purple before we can see it as such.

WE MAY LIVE inside a giant black hole.

A smallish collapsed star is very dense—a teaspoon of matter, if you could find a teaspoon strong enough to scoop it out, would weigh an enormous amount. But curiously, as black holes get larger, they become less dense. A black hole formed by the collapse of matter weighing about a hundred million times as much as our sun would only be as dense as water. "If you keep extrapolating, you will find that the density required to form a black hole with a mass equal to the mass of the observable universe would be roughly the same as the average density of matter in the universe," writes physicist Lawrence M. Krauss.[12]

However, if we do live inside a black hole, it is not one that strips us down to a few defining wavelengths. Whatever form of contraction exists at points in the universe, it still allows us multiple differentiated features of matter and energy, a spectrum of light. It allows islands of order to cohere.

Catastrophe is ordinary, inevitable, though we may not spend much of our lives in its immediate vicinity. And catastrophe is something that doesn't quite happen—the tragedy arrives, the monolithic pain grips us, yet the world does not come to an end.

Gather Ye Rosebuds

EIGHT

*Publishing a volume of verse is
like dropping a rose petal down
the Grand Canyon and waiting
for the echo.*[1]

How I laughed the first time I read this
maxim by Don Marquis! I've gone on
laughing at it all these years spent scattering such paper petals on the bank of
the North Saskatchewan and wondering
if they make any difference.

We grow lots of wild roses here; they scent our deep river valley in late June and stud it with rosehips that gleam and shrivel through the winter. Recently, we've added an array of hardy roses to our gardens, bred to withstand our brutal January temperatures and give us the frilled whorl of rose that we love from milder climates.

I love roses. I love poems. But the problem with writing poems is that it doesn't look like a spectacularly useful occupation. Other people feed and clothe the world, or govern it or reap investment from it. My husband goes every day to the inner city to an agency that helps people who have fallen over the social edge and whose voices can be heard echoing faintly in the city's life.

All my life, I've felt the guilt of not doing enough, of wanting to spend inordinate amounts of time staring out windows and rearranging words. I admire the active saints more than the contemplative ones, the people who attempt to make a difference in society and the world. What does writing a poem ever change? What difference does poetry ever make? What *is* change anyway?

AT THE BEGINNING of the last century, there was a general impression that science had reduced us to two rather discouraging accounts of change. The first was classical causality, the Newtonian picture of events being caused by forces acting on objects. Big things cause big changes that drown out little changes. Rose petals can't out-compete iron bars; the small ripples die away. The second idea that had become familiar and conventional was that of entropy. The Second Law of Thermodynamics holds that *all* ripples die away. A system constantly leaks energy; as time goes by, it becomes less and less ordered. The universe dies "not with a bang but a whimper," and it's all downhill from here.

Put together, these ideas combine into a deterministic view of change. If you know enough about the initial details, you can predict the outcome of every event in the universe. (Human beings weren't likely to manage this, of course, but in principle a demon or a god *could* have the necessary knowledge.) And science seemed to promise that we could figure out the main levers—we could reduce the world's myriad phenomena to a few key equations, the big causes that would effect the arc of change from past into future.

The puzzle of cause and effect has been a preoccupation of science from the beginning—not surprising, since causation is a default preoccupation of the human mind. We evolved to notice the patterns in our world that have predictive uses—"Hit the stone at this angle and a flake will break off." Aristotle's analysis of sufficient or necessary or antecedent causes is not merely an exercise in abstract reasoning—it's a process we constantly engage in. You can see it in the kinds of fine distinctions we make every day in language. Steven Pinker points out the "semantic fussiness" of verbs like "pour" or "fill," which may seem to mean pretty much the same thing.[2] You can pour water into a glass, or fill a glass with water. However, each verb differs in the aspect of the motion event that it cares about. To pour means to allow a liquid to move downward in a continuous stream. It has a certain laissez-faire quality—the water might go into the glass or all over the table. To fill means specifically to *cause* the glass to become full; water all over the table won't count.

Children crack the puzzle of learning syntax by learning how to class various cause-and-effect details of relationship. Is something an agent or an object? Is an action self-contained and complete or ongoing? Does it involve acting on something directly or as an intermediate force? Each class of verbs differs subtly in how it's to be used correctly (we say, "I drenched the table with water," but not, "I spilled the table with water"), and the differences relate to our underlying conceptualization of motion, change, causation.

But, though science picked up our natural human concern with what's going to happen next, we didn't much like its mechanistic stance that what's going to happen in the future *must* happen because of what happened in the past. When Newton's vision of the universe as a great ticking piece of clockwork had become de rigueur in the eighteenth century, Dr. Johnson commented caustically: "All theory is against the freedom of the will; all experience for it."[3]

The deterministic view of cause and effect seemed to define science in the early modern age and became one source of a pervasive resentment of the scientific. Physics seemed to rule out the kind of change that we experience in the world, the importance of small changes and accident. We all know that lives are more tossed about by detail than Newton's grand planets are. Instinctively we feel that you *can't* tell what the future is going to be; we are even reluctant to allow an all-seeing god to have that knowledge of what we *will* do. Determinism removes "the point" from our lives.

FOGGY MORNING, heavy frost on rooftops. The sun is a perfectly round dot of grey confetti beyond the thick mist. This sudden cold will put paid to a small flush of buds on my hardy

roses outside the back door. I had hoped for a last rose of
summer, a few more pink petals, but the weather has scooped
me. If only I had known there was going to be frost. I could
have covered the buds, saved the flowers. But the weather
report has failed me.

I'm fascinated by the Weather Network, that curious
phenomenon of our time—a whole television channel
dedicated to predicting the weather, twenty-four hours a day
and seven days a week. I confess to a weakness for watching
it, even when I could have a pretty good idea of what the day
is like just by sticking my head out the door. But I continually
check the weather maps—particularly the system maps that
show the jet stream flexing north and south like a skipping
rope flicked by a giant. Today the jet stream has sunk well
south of the 49th parallel, the skipping rope left to trail across
the floor of the country. We're on the side of the rope where
cold Arctic air sits over the pole and slips from one side of the
planet to another like a beanie on a bald pate.

The rose outside my door was developed by a poet. I didn't
know about that when I planted it—didn't even realize it had
been deliberately created or how closely its creator's existence
connected with mine. Georges Bugnet was a remarkable man
who died in a nursing home in an Edmonton suburb in 1981,
at the age of 102. That was just a few months before I arrived
in Alberta; the three decades during which we were both on
the planet were spent in different parts of the country. Our
lives didn't so much overlap as adjoin, like linked stamps on a
perforated sheet.

Georges had been born in France in 1879 and was educated
there for the priesthood. But then he had a major change in
his life—he married, and in 1905, he came to the far West

of Canada. He took up a homestead near Lac Majeau, eighty kilometres northwest of Edmonton, in the Glory Hills. There he raised his family of nine children, edited a French-language paper, published poems and novels, and bred useful crops for the weathers and winters of his northern home. He created this rose carefully, crossing a double wild rose from Russia with the Alberta single variety to produce the famous *Thérèse Bugnet*, named after his sister and now grown all over the world.

There is an article by him in *American Rose Annual* (1941), called "The Search for Total Hardiness." In it he talks about the patient process of pollinating his hybrids and trying out the resulting generations for hardiness and colour and scent. He was a practical man: "My idea is not so much to add new varieties to the gardens of the city-dweller as to produce tough stuff for my fellow farmers who have no time to coddle tender plants."

TODAY WE HAVE three new approaches to the idea of change, created through the careful cross-pollination that mathematicians and scientists do as they work on the sets of data provided by the natural world.

The first, most recent, approach is complexity theory, which tackles the question of how simple components can combine to become the world we have in front of us. Complexity theory goes head-to-head with the Second Law of Thermodynamics. It shows how order arises in systems as naturally as disorder. Perhaps entropy rules at the very largest scales of the universe, but in more local situations, structure tends to emerge— whether it's gravity pulling dust together to ignite into stars, or the molecular precursors of life emerging from the pre-biotic chemical soup, or synaptic connections in a brain adding up to perception, analysis, thought.

In complexity theory, a few key variables can be used to define the system and a few rules about how they are allowed to interact can lead to the emergence of enormously complex systems. Some combinations settle down to a repeated pattern like a regular decimal. But others lead to constantly changing states; they never settle down. In particular, simple rules for how simple elements take their next step can build complex adaptive systems that are capable of acquiring information from the environment, noticing regularities and acting on those patterns. Complex adaptive systems are in play everywhere, from the mammalian immune system to the way a child learns grammar.[4]

The second approach to change, born a little earlier in the twentieth century from Benoit Mandelbrot's work on fractal patterns in nature, is chaos theory—which is not about the shapeless void of myth but the way that ordinary systems like the weather transform. In chaos theory, a system will evolve into one of a number of possible states, but you can't pull out the variable that will predict an outcome. Tiny, tiny variations in the initial conditions escalate into bifurcations that become impossible to predict very far in advance. This is the world in which a butterfly's wing-flap might trigger a hurricane.

The third approach to understanding how one thing leads to another—the one that actually came earliest in the twentieth century—is quantum uncertainty. Subatomic particles are in a constant froth of change, colliding with each other, acquiring energy, shrugging off that energy again in the form of a photon or neutron or some other tiny packet. From this criss-cross of energy and matter, you can isolate one particle and look at the options open to it. You will have, statistically speaking, a very accurate idea of how likely any one "next state" will be. But, you can't tell which of those states that particular particle will change to. We can never collect enough information to determine that one particle's future. So there's a discontinuity, a moment that drops out of the cause-and-effect chain, when a random event happens.

Quantum uncertainty is a bit like quantizing time—or change, or causality—because it breaks transitions down to a smallest unit and says that this unit is not quite contiguous with the next. But because there are a limited number of states into which this randomness can evolve, you do get a smooth curve of causality. Patterns that are in effect predictable emerge from the billions and billions of interactions occurring at any instant.

The three approaches to change—complexity theory, chaos theory and uncertainty—overlap in subtle conceptual ways. For example, uncertainty and complexity theory share this much: there is something discrete that can be moved—blink—into a new configuration. In comparison, chaos theory is the most continuous of them all. There's no smallest limit to the decimals; the tiniest can ratchet up into change.

But the uncertainty principle also shares similarities with chaos. They are both kinds of change leading to a relatively small number of possible states. A helium nucleus won't decay into an elephant and a shoehorn. We know precisely what its options are and how likely it is to end up in any one of those states. Chaos theory, likewise, offers a system with a limited number of basins of attraction into which it can evolve. An atmospheric system can go from butterfly flap to hurricane; the temperature outside my door can range from −30°C to 30°C, depending on the time of year. But however long I wait, the wind will not whip up into a spaceship or the temperature drop to 0 degrees Kelvin.

It is complexity that can make elephants or roses out of quarks, because a complex adaptive mechanism is not a self-contained system. It is interacting with a larger environment.

SO WHAT CAN a poem change?

First of all, a poem changes poetry. Every new poem is another butterfly in an ancient, changeful, weather system. The esteemed Canadian poet, P.K. Page, finds an old Spanish form, the *glosa*, in which four lines from another poet are worked into your own poem. She writes a few *glosas* inspired by other poets. Within a decade, this particular innovation has caught the attention of poets across the country, and we're all trying our hand at *glosas*.

For a few decades, free verse held sway over poetry in this country. It formed a large, stable, high-pressure zone that— like a blue-skied prairie summer—seemed as though it would never change. It spawned a certain way of presenting poetry on the page that became all-pervasive to poets of the latter twentieth century. As Dennis Lee could write augustly at the turn of the millennium, the syntax-and-notation of free verse "may furnish the prosody of English poetry for centuries to come— as metrical prosody did in its time."[5]

But even as it seemed as though the hegemony was complete, ripples were forming at the edges. The *glosa*, for example, involves the use of end rhyme to tie the borrowed lines into a whole. Sonnets start appearing in literary magazines again. The shifting weather system of poetry picks up a grain from the past and amplifies it into a new pattern. It never stays the same.

However, I don't like to think of poetry as merely the patterns of a self-contained system. Such a system may have its fascinations and passionate fashions, but they do not assuage my guilt at gathering poetry's roses rather than tackling the thornier issues of society.

POETRY CHANGES BRAINS. More and more evidence accumulates to indicate that the tools and techniques of poetry are those with which we learn language in the first place.

"Motherese"—known patronizingly as babytalk—is the dialogue that human beings hold with infants, and its speech patterns are full of the devices used by poetry. As Ellen Dissanayake and David Miall point out, when we speak to very young children, we use patterns that are spontaneous and cross cultural boundaries.[6] They include all the dynamic features of poetry like pauses, rests, changes and stresses— loud and soft, fast and slow—to create rhythm that would be unnatural in everyday speech. In motherese, we break speech into lines of three to four seconds, which has been shown to be a universal characteristic of poetry. We use parallelisms and hyperbole, mock seriousness and exaggeration: "WHAT a BIG YAWN!"

Infants don't know what we are saying with this distinctive patterning, but it attracts and holds their attention. (And what does any poet want but to capture attention with her words?) We use all the resources of speech sound—intonation, rhythm, phonetic colour—to capture or re-engage babies in what we are saying, creating an order that they can perceive to be structured and systematic.

These early interactions with speech patterning are extremely important in a child's development. Researchers have found connections with everything from an infant's homeostatic balance to its ability to function socially. Motherese teaches us the emotional and behavioural co-ordination that is our human endowment: "...it is by means of intrinsic poetic ('fore-grounded' or 'aesthetic') features that the coordinations of mutuality are supported and reinforced."[7]

Poetry does not only help babies to learn language. A great deal of research shows how important rhyme and other language play are to the development of reading and literacy. Knowing nursery rhymes by heart at the age of three is a good predictor of reading development at age six.[8] There's a correlation between a five-year-old being able to figure the "odd man out" in combinations like cat/fit/pat and her reading and spelling progress at the age of nine.

So the processes of poetry are central to exploring the potential regularities of language. A child learns, "This combination of sounds belongs together; it can be treated as a unit and moved around. This combination makes her pay attention to me." In effect, a child is learning not only language but the powerful process of inductive logic that underpins our ability to reason about the world.

We have privileged an opposite way of thinking, *deductive* logic, since the days of ancient Greece. The search for a chain of steps from an absolutely certain starting point to an absolutely inevitable conclusion has been the goal of thinkers for centuries, *the* route to knowledge. Induction, on the other hand, is the conviction that the sun will rise tomorrow just because it always has—an assumption that can be fatally flawed if you're the turkey expecting dinner from a friendly farmer on the day before Thanksgiving. "Induction is not to be trusted," said many scientists. "We need firmer ground."

But inductive reasoning is, in its own way, more powerful than deduction. The great deductive towers of cause and effect can be just as untrustworthy—get their starting points wrong and the edifice collapses. The capacity for induction, by contrast, is robustly constructed out of myriad interactions and possible-but-not-certain causality; it deals with systems that fall more-or-less reliably into typical basins of attraction.

All animals evolve to recognize and respond to such patterns. ("Most of the time that pattern of two dark spots is a predator watching me, so hustle out of here.") Humans are even more intensively exposed to induction as we learn language, that massive stream of changing data that comes in at a baby and has some puzzling connection with the world. Over the first few years of her life, she will learn not rigid rules but a flexible set of patterns that allow her to use language fluently and correctly. The patterns of poetry become her theorems.

MY EARLIEST MEMORY involves language. I am sitting on the floor, under the table in a kitchen. I think it is my grandmother's kitchen—the world beyond this space is undefined in my memory. But I am aware of the chair legs at my back and the bottom of the table like a roof over my head. I am thinking, "My name is Alice." I'm aware that this is a piece of information that I didn't always have, that there was a time, fairly recent, when I did *not* know my name was Alice—that a cave lies behind me from which I am now looking out.

I might have been about two and a half, judging from the relative size of me and the height of the chair seat behind me, and the table overhead. Clearly I had enough of language and syntax to have an understanding of "before" and "now." It was a moment of quiet eureka, of deep satisfaction and wonder. I had a sense of liking the sounds of my name—the roundness of the "a" and the pleasant linger of the "l." Something of importance had been presented to me, a gift as complex and layered as a hybrid rose.

POETRY CHANGES BRAINS QUITE PHYSICALLY.

"The utmost of ambition is to lodge a few poems where they will be hard to get rid of, to lodge a few irreducible bits..." wrote Robert Frost.[9] Poetry has been called a "technology for memory."[10] That process of lodging a few memorable bits in someone else's brain requires a literal change. In the process of making memories, the brain's neurons lay down small calcium spines that enable a synaptic connection to be made and made again more easily all the time. An indelibly remembered poem is a flower growing from thorns.

PERHAPS THE MOST IMPORTANT changes made by a poem involve hearts. Or perhaps it is not so much a matter of changing as crystallizing emotions already there but unformed. It's as though a poem can help us to recognize ourselves. I was always particularly susceptible to a certain kind of romantic emotion when I was young, to poems that seemed to promise me a wild loneliness, the possibility of living alone in a bee-loud glade. I still have two little notebooks with black moiré covers in which I would copy out poems like "The Song my Paddle Sings," or "When I Set Out for Lyonesse," or Walter de la Mare's "Nod":

> *His are the quiet steeps of dreamland,*
> *The waters of no more pain*
> *And his ram-bell rings 'neath an arch of stars*
> *"Rest, rest, and rest again."*

I wrote the verses out with coloured ink bought at Woolworth's, turquoise and gold and a red that has faded now to old rose. Perhaps it was my way of surviving life in a cramped apartment, to know that words could create empty spaces.

For years, I thought of this taste as childish, something to be suppressed and disowned. In university, I read T.S. Eliot's decree: "Poetry is not a turning loose of emotion but an escape from emotion; it is not the expression of personality, but an escape from personality."[11] Or there was Charles Olson praising his approach of "objectism" as "getting rid of the lyrical inter-ference of the individual as ego, of the 'subject' and his soul."[12] I had the general impression that I needed to get rid of my lyrical ego quick-smart if I was going to be a modern poet.

But, as David Miall points out, emotional response is central to how we navigate literary works. Emotion is how we remember the subunits in the text, how we stretch the little window of working memory wide enough to hold something as long as *The Iliad* in our minds. Emotion is what we put into a work as writers and what we log into as readers.

This is certainly not a new thought; there is a long tradition of poetry (and art in general) being associated with emotion, while science is consigned to conveying cold information. What has changed is a sense that emotion is neither a poor cousin to logic nor a grandly stated but fuzzily conceived superior faculty. The cognitive research is telling us that we learn—i.e., change our brains—by how we react emotionally—i.e., how we pay attention with our hearts.

Nor is poetry only about the solitary dreamy emotions. It also energizes anger and claims power. That has been the legacy of hip hop—a reclaiming of agency by young people using words. It is significant that hip hop has been marked by the heavy rhythm and social bonding; the rapper in front of a live audience moving to the beat, a physical engagement that energizes the whole group.

YOU NEVER KNOW how a poem might change things. I got an e-mail once, asking if I was the author of "Puce Fairy Book"— one of those poems that finds its way into a school anthology and haunts its author forever after. The inquirer said she had found the poem pasted to the door of a girl's closet. I wrote back saying, yes, I was responsible for it.

"Oh, good," came the response, adding that she intended to give a copy to her husband for Christmas. "Not that he'll understand it," she wrote darkly. "If he understood it, I wouldn't have to give it to him." I wondered if "Puce Fairy Book" might become the first Canadian poem cited in a divorce case.

More inspiring are the cases where poetry has directly helped people survive hardship—by knowing a poem by heart and holding onto it as a talisman. My friend Shirley Serviss works as a poet on the wards at the University of Alberta Hospital and has many such stories of people for whom poetry matters in extreme situations. She writes,

I want this poem to be better
for you than cod liver oil
and easier to swallow…[13]

Joseph Brodsky, in his address on becoming the US poet laureate, tells of seeing a copy of one of Robert Frost's books, which opened by itself onto the page with the poem "Happiness Makes Up in Height for What It Lacks in Length." Across the page went a huge size-twelve imprint of a soldier's boot. The front page of the book bore the stamp "STALAG #3B."

Brodsky comments, "Now there is a case of a book of poems finding its reader. All it had to do was be around. Otherwise it couldn't be stepped on, let alone picked up."[14]

WHY SHOULD I CARE if poetry changes anything? A poem isn't a screwdriver. Surely it is created as an object whose purpose is to be itself, not, god help us, a "change agent."

This is such a pervasive argument to justify art in our time. It has been the artist's response to the scientific vision of change as Newtonian determinism, the view that the future is caused by the big things of the present. What poets don't seem to have noticed is that, in making it, they base their argument on the very same view of change. They make the same assumptions about how causality works as the scientists of the nineteenth century did and say, "The hell with that. We're not going to play that game. Our art stands aside from your clockwork universe."

What artists haven't really noticed is that the scientific picture of cause and effect has changed profoundly. It no longer presents the future-directed narrative we found so dubious. It says we *can* understand the steps that got us here because things are connected, but we *cannot* know what the future will be at any great remove because too many things can change it.

All I ask for poetry is that it be connected to the world, that it not stand aside. A poem may be a very small decimal point in the constantly evolving adaptive system of human culture. However, as with chaos theory, no decimal place is too tiny to count.

ONE OF THE TRICKS that hardy roses use to survive bitter winters is the concentration of sugar.[15] Plant tissue is made up of two kinds of cells. Structural cells are basically fluid-filled sacs—membranes that surround water and nothing else. The second kind includes the engine of the living cell—the nucleus and the delicate chemical machinery of its metabolism. Freezing the first type of cell does no real damage to the plant. However, freezing the cells that allow the plant to function will kill it. Ice crystals can tear apart the membranes that surround the cell, spilling its living contents out, or tear apart the protein molecules within it.

Regular water freezes at the well-known temperature of 0°C. However, water that is laced with other substances, such as sugar, will solidify at lower temperatures. During the cooling darkening days of autumn, hardy roses build up higher concentration of sugar in their cells, a slowly thickening syrup to help them avoid freezing. The membranes around the cells

become more flexible and also allow water to transfer in and
out of the cells more easily. This helps the cells withstand
periods of extreme dehydration. It's a way of facing change.
This art of concentrating sugar has been forgotten by many
of our coddled roses. But it is a wonderful metaphor for the
survivability of roses and poems in the wild. Poetry survives by
concentrating language and adapting to local conditions.

GEORGES BUGNET'S LITERARY LEGACY is local, largely
unknown to most of the world outside this region near his Glory
Hills. His novels are set around here, and treat Aboriginal
people and the Riel Rebellion with respect. Most of all, I like
the details of forest and animal life, muskeg and marsh hay
that make me feel as though I'm reading a work by a friend. To
give you a flavour, I've translated a couple of stanzas from one
of his poems, "Le coyote," published in his book, *Voix de la
solitude*:

> *Under the fall of night, out of the rough brush,*
> *the coyote comes silently.*
> *He lengthens his neck, watches, sniffs the wind.*
> *He points his ears and quivers....*
> *He traces a narrow line in the snow.*
> *No sound startles the dreamy silences.*
> *He glides, living shadow, in the glimmer of death,*
> *and his step quiets itself in the quilted ground.*[16]

It is a sombre piece—the coyote dies of strychnine poisoning,
and priesthood-trained Georges cannot help drawing a
religious comparison with the human inability to resist

temptation. However, it is the details about the animal itself that I love. There is even a little footnote in the book to explain how the characteristic trail of the coyote forms a narrow line. That, more than anything, reminds me that coyotes had probably found their way into very few poems by French literati of the early twentieth century.

Georges Bugnet did not care much for the literary currents of his time. In the preface to *Voix de la solitude*, he talks about the influence of Verlaine and the efforts of modernist innovators: "I have little doubt this pleases an elite...However,... most readers are cast afloat on unfamiliar waves." His attitude can be interpreted either as the stubborn conservatism of the provinces or thoughtful independence from fashion.

> *don t cuss the climate*
> *it probably doesn t like you*
> *any better*
> *than you like it*[17]

SO WROTE Don Marquis' literary cockroach, Archy, who didn't spend a lot of time watching the Weather Network. Archy was too busy with the exhausting effort of trying to get words on a page, throwing himself headfirst onto the typewriter keys. He didn't undergo the angst of wondering whether the effort was worth his while.

I will take heart from that. And from the way that hybridizing roses seems so wonderful a metaphor for poetry itself. Something is patiently pulled together from roots in different parts of the world. Something can flower into a new form from modest ancestors. Something is always "a search for total hardiness."

As Georges the rose breeder wrote in that 1941 article: "At the present time no one can tell how long will be this work. Fair success might come out of just one mating; more likely it will require a great many....I may be able to hasten the hour of victory. Be that as it may, one can hardly wish, in these years of terrible wars, for a more pleasant 'job' than that of endeavouring to obtain, for the benefit of man, new favours, new clean and long-lasting gifts."

NINE Motion

I'm working soap into my knuckles, grimy
from my battle with the bellflowers.
Creeping bellflower, to be precise, or
Canterbury bells as they're sometimes
known—names that imply a demure and
delicate quality. But far from abiding shyly
as soft purple chimes in a corner, creeping
bellflower is what the garden books call
an "invasive" species. Invasive! That's as
inadequate as calling it "creeping." This is
the vegetable equivalent of Genghis
Khan's army. It springs up, shaking its
purple spears all summer and scattering

seed, determined to repopulate conquered regions with its offspring. And its roots dig in, the pervasive troops of colonization. We think of plants as motionless stay-at-homes, but they really are out to cover as much territory as they can.

I'm really not much of a gardener. Most of the time, the strip of perennial plants at the edge of my yard looks as though it's reverting to tall grass prairie, and I let poplar shoots sprout on the lawn until it's on the verge of going back to aspen parkland. But every so often, vegetation goes just too far over the border and some territorial imperative makes me take a stand. So for the past week, I've been digging up several yards of earth, tracking resourceful roots through their underground maze and liberating the other plants who would like to live there.

Creeping bellflower roots have an odd fascination for me. Their criss-crossing tangle of threads connect at nodes where a triumvirate of white fingers pushes down into the soil to fatten themselves. Leave the merest fragment in the ground and it will catch its breath for only moments before resuming its pronged burrowing. I feel victorious when my spade gets deep enough to let me lever out a whole, fleshy, pointed cone. "Gotcha!" I crow and wave it like a pennant.

As I try to get clay out from under my fingernails, I find myself thinking again about how different the body plans of plants and animals are. Plants are essentially linear.[1] They are collections of tubes joined end-to-end, made of cells that are adapted to being conduits. They grow from tiny nubs of tissue at the tips of those bundles called meristems, from which a stream of cells pours backward like a fountain. The meristem at the end of each root has a navigation system that can pick its way around stones and sense gravity. And plants can just keep expanding, using the trick of vegetative reproduction. The largest organism in the world may be a stand of aspen in Utah that has colonized an area of more than one hundred acres.[2] Its single root system thrusts up tens of thousands of trunks, all of them clones of one another.

Animals begin from a small nub of proliferating cells as well, but as embryos we build more or less from the inside out. Where plants get bigger by expanding the size of individual cells, animals grow tissue by continuing to subdivide cells. They get to a certain size and stop. When animals want to change location, they have to pick up and go as a whole; they don't have the option of sprouting from bits of themselves.

THE POEM I was working on earlier today seemed to be following a vegetative body plan, which is probably what drove me outside to attack the bellflower. Sometimes you sit down with a piece of paper; you're not going to wait for the animal surge of inspiration, goddammit, you're going to write something. You let words take a random walk—or what mathematicians call a "drunkard's path," a step in any old direction followed by an equally random step in another. It's a model that can carry you surprisingly long distances based on many small decisions.

Maybe a chance encounter happens between "cloud" and "custard" that makes you look out the window at the sky and note its curdled lumpiness. "Cloud a dun custard," you scribble and wonder where that exploring root tip might lead you. Nowhere much, you decide, and head back to the zero point outside the tavern door. Which way this time?

Some writers love this way of proceeding, don't want to know where they're headed in advance. Frankly, it drives me nuts. I like a plan, a map, a destination. Instead of shaping up like a drunkard's path, I'd rather have a poem grow like the border of the Mandelbrot set, with a general recognizable shape that becomes ever more elaborate as I work on it. Yes, the poem I'm working on may head off in an unexpected direction, but I then immediately want to recalibrate, to see if I can make this fit with the original plan.

However, sometimes you do just have to get in there and follow the root system, yank up whatever comes next. After all, the body plan of flowering plants does come to satisfying conclusions. As the genes of a meristem build a shoot, they issue a map of signals—a map that is being made at the same time as it is being read. When the conditions are right, the meristem changes character and makes a loop, a whorl of four closed rings instead of a continuing spiral. It forms a flower and stops. It has reached its destination, just as the most randomly proceeding poem will.

"YOUR POEMS HAVE A GREAT DEAL OF CLOSURE," a young creative-writing grad told me once during a master class that I'd signed up for. She added helpfully that "this isn't fashionable in Canadian poetry these days."

I refrained from telling her to teach her grandmother to suck eggs, and thought about the odd modern fad for leaving the pallid little legs of a short story or poem to kick in the wind. "Is *that* the end?" the reader wonders, turning the page to see if there is something she missed. For a while it felt as though contemporary writing had been taken over entirely by the vegetable body plan.

Of course, this approach can be used wonderfully in literature; we don't want everything tied down with knots.

In *Alice in Wonderland*, one event follows the other in the random way that episodes in a dream emerge out of what has just happened before but with no overall destination in view. However, much as I love Wonderland, I prefer *Alice through the Looking Glass*, because it actually has a plot. Alice begins as a pawn on the dream-chessboard and ends up as a queen. However arbitrary the intervening episodes are, they come to a natural and satisfying conclusion: "Welcome Queen Alice with ninety times nine."

The two alternating approaches have been followed for centuries. *The Thousand and One Nights*, for example, is vegetable. Its stories unfold one after the other from the meristem of Scheherazade's voice, as she tells tale after tale, night after night to intrigue the king and save her life. There is no final episode, no reprieve, no conclusion to the process— the original version of the work has no final reconciliation of the king and the taleteller. It stops as abruptly as the most postmodern of constructions.

On the other hand, Geoffrey Chaucer's *Canterbury Tales* does have an end in sight, even though his collection of stories can seem as random a work of bricolage as Scheherazade's. He sets his purpose out clearly in the prologue: his pilgrims on the road to Canterbury will each tell several stories suited to their diverse ranks and personalities. Chaucer did not live to complete his massive project—he died in somewhat obscure circumstances, caught perhaps in the fallout from a political regime change.[3] But even though his project exists in clumps and fragments, we can see its overall intent, which is nothing less than a portrait of his whole society. And he knew how it would end. We do have the concluding "The Parson's Tale," in which Chaucer abandons his sparkling lines of poetry for a very lengthy prose meditation (puzzling to many commentators) about a virtuous life. There is a lovely image in the prologue to "The Parson's Tale" that puts it in context. The troupe of pilgrims is coming close to the edge of a village, and Chaucer-the-narrator notices the length of the shadows in the declining sun. They are about eleven feet long, or about twice as long as he is tall:

> *Of swich feet as my lengthe parted were*
> *in six feet equal of proporcioun.*

There is something of mortality in the image of the shadow on the ground, the six-foot length of the grave. The Host points out that there is one last teller to be heard from and commands the Parson—a good man, modest and kindly, on the edges of the group throughout the enterprise—to unbuckle his bag and out with it. "Say what you please, and we will gladly hear," says the Host, and the Parson begins his long sermon on what should move a man to contrition. Chaucer's pilgrims are meant to get not only to Canterbury in their merry, worldly, bickering band; they are also to get safely home again, with a greater understanding of what's important about life and death.

THE DIVISION between plant and animal is ancient. Our lineages have been travelling their separate-but-connected paths for two billion years, and the difference between us comes down to different needs for motion.[4]

At some early point, small, green, bacterial cells developed the trick of harvesting the energy of sunlight to break down water and carbon dioxide, substances that were plentiful in their ocean environment. They were in close living quarters with other cells that could use these by-products, and the relationship became so useful to both that the small green bacteria just moved inside the walls and went on with their task of capturing photons and releasing chemical changes. They gave up the genes they didn't need, since functions like building conduits and competing for sunlight could be carried on by the larger cell inside which they lived. Today, every plant cell contains these remnant bacteria as chloroplasts. We can still make out the ghosts of their old walls as membranes around their minute bundles of DNA and activity.

These tiny factories began pumping out vast quantities of oxygen into the planet's changing atmosphere, making another possible source of energy available to the seething, twining, replicating molecules that filled the oceans. Oxygen is a violently reactive element—it leaps onto other atoms and locks arms with them, spitting out the leftover bits that don't fit into the new configuration, freeing them to become part of other chemical compounds.

The hard work of the chloroplasts meant the free molecules of oxygen and carbon that had been so conveniently broken out of carbon dioxide were now available to be used as fuel in an entirely different metabolic process. A different kind of bacterium developed the new chemical trick of using cast-off oxygen for another set of chemical reactions. This pioneer was then also welcomed inside a larger cell, where it, too, gave up the functions it didn't need any longer and focused on churning out energy. These bacteria became the mitochondria that nest in every animal cell—rings of separate DNA inside their own ghost-walls that drive our metabolism.

Here's where movement enters in. Chloroplasts can do their work wherever there's a photon of sunlight, and sunlight falls in fairly predictable patterns around the planet. All you have to do is show up and hold your arms out. Mitochondria, on the other hand, need plants. They need the oxygen they give off and their carbon compounds. An organism that consumes all the plant cells in the neighbourhood has to move on, by drift or determination, to find alternative sources. From the earliest days of animal evolution, we've had to follow plants around— even if we don't eat them directly, we eat other animals that do. We are appetite in motion.

WE STAGGER AROUND when we're drunk because the activity of our cerebellum has been suppressed by alcohol.[5] The cerebellum is a smallish lump tucked under the neocortex, just at the top of the brain stem. When it was first identified, scientists thought it was merely required for the relatively simple task of co-ordinating movement.

In fact, a brain is first and foremost a system for initiating and maintaining movement, which takes co-ordination of thousands of muscles hooked to bone so that they maintain balance, respond to incoming information and lever the body into new positions as needed. I watch a dog catch a Frisbee and I'm awestruck by the complexity of the functions needed to perform that twisting leap, to snap at air and come down with a plastic saucer. The addition of judgement and abstract reasoning seems like a relatively modest addition to such a gloriously effective system.

The human cerebellum is an elaboration of a very ancient organ. An early version of it appeared in fish, connected to the inner ear and used for balance. It developed further in amphibians and reptiles, to control posture by controlling muscle tone, keeping muscle fibres in tension and releasing them as required to enable movement. The cerebellum became even bigger in primates and humans. Today, the cerebellum helps fine-tune movements produced elsewhere in the brain, making them so fluid and harmonious that we are not even aware of how complicated it is to walk down a set of stairs. It is a major part of how we store procedural memory—the way we remember how to ride a bike or say a word.

The cerebellum is still small compared with the balloon of the neocortex, weighing only 10 per cent of the rest of the brain. However, it contains more than half our neurons.[6] It is massively interconnected with the neocortex, particularly with the emotional centres. Motion increasingly came to involve *e-motion*—those feelings that "move out" and take us with them.

"MY BRAIN IS LIKE A DOG'S," my father said one day, when Alzheimer's was already well progressed. "It just follows whatever is in front of it." It was a sadly insightful remark. One of the first things that Alzheimer's does is erode the ability for long-term planning, because that takes long-term memory. If you cannot remember what your intentions are, you cannot complete them.

Earlier still in the progress of the disease, he had become lost one day downtown, separated from my mother by some swirl of pedestrian motion on a busy street. Hours and hours of panic and phone calls later, my mother opened the door of their apartment to find him standing there.

"Bill, how did you do it?" she asked later, after she got him in the door and settled down with tea and biscuits. "I just came the way the car would drive," he answered. His mind could no longer make a map of the city or recognize the relationships among the named streets he was walking on. He couldn't work out that he was close to a subway station or a bus route. But he found a familiar corner and a kind of motor memory kicked in. He did not have an overall picture of where he was going, but at each corner he could make one small non-random decision after another, based on long years of driving that same route, to find his way to that place he remembered as home.

THIS YEAR of my bellflower battle is also one of the years when thousands of pelicans came home to Lake Eyre.[7]

Australia is like a cake that has fallen slightly—a rim of faintly bumpy crust around a depression in the centre. Lake Eyre sits at the lowest point of an immense drainage basin in the middle of that baked surface, about fifteen metres below sea level. Four main rivers run *into* it, rather than out toward the fringes of sea, and their water collects and evaporates, leaving a hard-crusted icing of huge saltpans.

There's usually a little water in Lake Eyre. But every thirty years or so, when the rainfall conditions are right, the rivers fill and bring their fresh fluid freight to spread in a vast shimmer over more than 9,000 square kilometres. The lake rushes instantly into life, full of micro-crustaceans and fish. Plants leap into leaf and bloom. And from hundreds of kilometres away, white pelicans leave their usual breeding grounds at the coast, sensing the coming flood less than twenty-four hours after the first trickles of water seethe through the salt crust. In this suddenly fertile paradise, thousands of birds will be born. Then the land begins to dry again; evaporation makes the water saltier and saltier; and the pelicans head back to the coast to live there for years or decades until the next miraculous flooding.

The mystery is: how do the birds know when to return? How does a pelican sitting on the coast know to head for the desert? What cues is the environment sending and how do they read them? A pelican's life span can stretch as wide as its wingspan: some of them live for thirty years or more. Biologists have speculated that the oldest birds might somehow remember the signals and lead the way back to the shining territory where they were born. But what can pelicans remember?

Animal migration is one of the great wonders of this moving world. Creatures from butterflies to blue whales can cover half a planet in their annual search-and-return to the places where they reproduce their kind, using an array of techniques to find their way.

Homing pigeons, for example, rely on an elegant combination of abilities to find their way back home after they've been dropped off a thousand miles away.[8] They can use the sun's position to orient themselves and have an internal clock and compass to judge their location in relation to the sun's movement through the day. They have a second compass that senses Earth's magnetic field, as well as minute amounts of magnetic iron ore in the neural cells of their upper beaks. They can hear sounds as low as one-tenth of a herz—far below the lowest rumble of an organ pipe—and researchers speculate this may enable them to hear winds blowing across mountain ranges thousands of miles away. They seem to use more ordinary cues from scent and vision to navigate the last few miles home. Finally, they are very receptive to barometric pressure, and their feet can sense minute vibrations. A pigeon is in contact with its planet's environment from the largest to the smallest scales.

HUMANS DON'T HAVE this impressive arsenal of orienteering techniques. You can't put us in a dark box, drive us to an unfamiliar destination and expect us to make it back home entirely on our own internal resources. We generally need compasses and maps, signs and passers-by to direct us. Yet our species burst out of the box and made its way around the entire planet within a remarkably short time.

The earliest physical remains of *Homo sapiens sapiens*—the OMO skulls found in Ethiopia—have been dated to 200,000 years ago.[9] Genetic evidence gives us a roughly similar age range for our species. The separate ring of DNA found in mitochondria indicates that everyone on the planet can trace this particular piece of genetic heritage (which goes from mother to daughter) to a female ancestor who lived approximately 190,000 years ago.[10] The OMO skulls belonged to people who were essentially modern humans. However, if the people who owned them were entirely modern, for many thousands of years they didn't seem to be doing anything spectacularly different with their human brains. They made the same stone tools and lived much the same life as their immediate ancestors had.

About 125,000 years ago, we start to find evidence of symbolic behaviour. Pencils of red ochre show up in the Klasies River Mouth Cave at the southern tip of Africa. At nearby sites, a child's bones have been stained red with ochre—a manipulation of objects for some purpose we can barely imagine but which imply that death is being thought of in a different way, one we might call metaphorical.

Tens of thousands of years go by. Up to this time, *Homo sapiens sapiens* had stayed within Africa, although earlier relatives had wandered the world as far afield as China. Then, around 70,000 years ago, genetic evidence suggests that our species went through a population bottleneck that reduced us to perhaps 15,000 individuals.[11] Such a bottleneck might have been caused by environmental conditions such as prolonged drought in East Africa or the volcanic winter caused by a massive eruption at Toba in Sumatra. Whatever the cause, shortly after hitting this point in our story, modern humans turn up everywhere.[12] Within the next 20,000 years, we were living in China and Papua New Guinea—and perhaps were even lighting fires in Brazil.[13] By 50,000 years ago, we had certainly reached as far as Australia (a journey for which we needed seaworthy boats) and were living at Lake Mungo in the southeast corner of that huge country. Today, Mungo is another swathe of dry sand in the Australian interior, edged by enormous crescent-shaped dunes. Back then, the dunes were the shoreline for another lavish ecosystem. Perhaps the people who had found their way there so many thousands of miles from Africa also watched white pelicans arrive to nest.

The puzzle is, what caused our "great leap forward," as it has been called. Did we simply achieve some sort of cultural critical mass? Or did some final genetic tweak suddenly create the ability to represent animals on cave walls or carvings on bone. If such a tweak did happen, it might have occurred in that small group of survivors who made it through the bottleneck. However, if such a sudden increase in brain capacity took place at the bottleneck point, we have to explain the fact that we were using red ochre to colour things 50,000 years earlier, as well as fishing with bone points and using many other small incremental improvements to help us hunt and survive in groups.[14]

This may well be a false problem, based on the assumption that cultural activities always leave solid objects for anthropologists to dig up. We used to think we could only go as far back in time as we could find historical documents, records, pictographs, bones, stone tools. It seemed the deep past was lost to us unless we had these obvious objects to build reconstructions on. But lack of fossils no longer means a dead-end for exploring the past. Now we are learning to read the record of genes. The transparent swirls of DNA that geneticists can isolate in their test-tube vials are now evidence as strong and clear as any stone tool. Through it, we learn the record of our comings and goings.

In human culture, our first symbolic artifacts are as likely to have been made with the swirl of words as in any less tractable medium. I suspect that our first such artifact would have been a map—a map made, not of diagrams and pictures, but of words. "You will get to the fruit trees by crossing the river and going past that small hill."

OF COURSE recovering the history of language is just as fraught as any other aspect of symbolic behaviour. When did *that* start? What led up to it? What needed to change in our brains to allow it?

Language was almost certainly the turning point in modern human development. And language has an all-or-nothing quality that makes it hard to imagine a partial version. It doesn't take thousands of years for a pidgin—the kind of speech used when speakers of two different languages have to negotiate trade relationships—to develop into a creole, with grammatical relationships that allow speakers to say anything they want about anything, past or present. The development of a creole takes only a generation or two. The same thing has happened with sign languages around the world. Deaf children exposed to halting iconic gestures of a pidgin signing system spontaneously created the rich referential patterns of American Sign Language.

It seems to be an unstoppable drive in humans. As soon as you have the idea of using combinations of sounds as the name for objects, and the idea of different sets of sounds to indicate relations between them, you get the whole language shebang in a mad clatter. What would be the use of speech that didn't allow you do it all?

However, from a physical, cognitive point of view, language is far too big a structure be acquired suddenly. A long development of protolanguage had to occur first, and we share its roots with other creatures. All mammals make communicative noises using their throats and larynxes. Our much-vaunted abilities of abstraction—categorizing objects and associating them with some combination of sounds—also has roots in the animal world. Categorization is part of pattern recognition from the earliest development of perception. Many animals and birds are capable of counting small numbers, which essentially requires the ability to group a set of stimuli as objects in a category and keep track of them. Vervet monkeys use specific calls to warn their fellows of predators. The calls themselves seem to be innate, hard-wired, but young monkeys do have to learn to use them properly. They will apply the "eagle" cry to crows or even falling leaves, until they refine the category of "moving-thing-of-a-certain-size-in-the-sky" to the specific creature that is their most dangerous predator.[15] Even recursion, which requires the recombination of smaller units into larger ones and seems to be fundamental to language, has conceptual roots in the minds of other animals. Any dominance hierarchy requires its members to keep track of which individuals and subgroups are subordinate to others. Living in groups requires a lot more memory.

In early hominids, capacity for language very likely evolved along with music and gesture under the intense pressures of socialization. For many millions of years, our predecessors had no problem with the immensely important function of communicating how they felt. The leap into true language seems to come down to the basic process of matching a word—a noise—to a concept for the purposes of exchanging, not emotion, but information. There is a great deal of debate over what might have propelled that leap. Is there one central capacity that suddenly emerged—such as the ability to create new symbols and learn them, or the ability to use rules of syntax to combine them?[16] Or was there a slow and gradual development of many interrelated brain systems, such as memory?

In spite of language's all-or-nothing quality today, our hominid ancestors could have gone on for many millennia using a simple pidgin to provide directions and exchange bits of immediate knowledge. Even a pidgin is a very useful tool, for all its limited grammar and combinatorial abilities. In modern times, there is a suggestive commonality to the many pidgins that have developed around the world; they share features regardless of what languages have been combined to make them.[17] For instance, pidgin speech focuses on the here and now, the tangible: the first thing that we need to talk about. All pidgins tend to indicate semantic relationships using the order of words rather than by modifying the words themselves (as we make plurals or verb tenses in English), even if their parent languages used morpheme changes to create syntax.

The question is, when would we have made the leap as a species from pidgin abilities to full language? When did the ability to use syntactical grammar come to be? The transparent swirls in test tubes may help answer this opaque question. Geneticists have identified one particular gene, FOXP2, which has mutated into a slightly different form from the version carried by our close relatives, the chimpanzees. When the gene malfunctions, it affects a range of language-related functions, including the ability to make the muscle movements associated with speech and, intriguingly, syntax.

At first, scientists thought that the FOXP2 mutation happened about 200,000 years ago—a nice fit with the appearance of the earliest *Homo sapiens sapiens*. But more recent analysis indicates that the change happened a good deal earlier.[18] In fact, it happened early enough to have occurred in the common ancestor of our species and Neanderthals. (The Neanderthal people were our close cousins, but not, according to genetic evidence, our ancestors. They did not seem to paint things with ochre or bury their dead with ritual, but they could survive the cold climate of Europe, which would have required an ability to manipulate the world's objects purposefully and communally. I expect they had lullaby and praise songs, war songs, names for themselves and terms of endearment—the language of the moment.)

FOXP2 is not *the* language gene, of course, or even the grammar gene. The genetic structures that facilitate our acquisition of speech are numerous and complex, and researchers will be disentangling them for years. But I do find it intriguing that genes *can* affect our ability to acquire syntax—in other words, to carry out the linguistic manipulations that go beyond simply specifying objects in the immediate environment and current relationships between them. People with damage to FOXP2 have problems with the kinds of inflection used for verb tense. A different kind of genetic glitch affects our ability to use a rule to form plurals. These are the kinds of abilities that allow us to move beyond pidgin speech, and the fact that they can be disrupted by a genetic malfunction gives us some hope that we can disentangle not only speech problems suffered by individuals today but the mysterious past of language.

IF THE BRAINS within the OMO skulls were fully modern but not from a people who made the kind of rock carvings or other artifacts that we think of as symbolic, what might they have been doing instead with their new capacities? I expect the first really useful thing would have been an ability to make maps, to pass along information on how to go and get back, even past the age of the oldest member of the tribe. This would have been an enormous advantage for the species that possessed it.

A map is a highly abstract object. It requires the ability for nested concepts that put ideas into recursive kinds of patterns: "travel for two days until you reach a pass that runs beside a river between two peaks." It needs to be flexible and portable. When you reach a place of abundance, you can settle down and make solid objects that embody knowledge, but as a wanderer through marginal territory, you have to travel light.

A map also has to be highly reliable, something that can be passed from individual to individual with a high degree of exactitude. Today's map is a visual object, but I think the first maps would have had to be poems—structured compositions in language that could be remembered and passed on.

This is only speculation on my part, among all the speculations about what set our species apart and on our particular course. But a few shreds of suggestive evidence support it. One is the tradition of songlines in the culture of Aboriginal Australians, who feel that their totemic ancestors had scattered trails of words and musical notes along the line of their footprints, and these dreaming tracks lie over the land and its features as ways of communication among far-flung tribes.[19] The songlines are similar to the landscape-reading traditions of the Inuit, who have passed along a vast geographic knowledge through many generations by oral means, without the use of maps or any other written documentation.[20]

And then there is the widespread hero myth of quest and return, found in cultures around the world. So widespread an artifact surely demonstrates the importance of individuals going forth and coming back with benefits to the tribe.

It hardly matters whether the final tweaks to language that enabled faithful transition of maps occurred in the OMO skulls or in the post-bottleneck population of 70,000 years back—though I think it likely the OMO people were making verbal maps. Even before we started our feverish drive to the far ends of the planet, we needed maps. Much earlier, we had trade routes within Africa that extended for hundreds of kilometres. The ability to transfer such information about getting there and back again would have been essential to a nomadic species that did not have the navigational abilities of pigeons or pelicans.

WHY DO I—we—like closure so much? Creative-writing grads aside, people seem to be happier with a work of literature they feel has been satisfactorily concluded. I can't make this statement categorically; I'm not familiar enough with the literature of other languages. (Perhaps this would be an interesting research project.) And individuals vary widely in their liking for open-endedness vs. completion. However, for argument's sake, let's say a preference for satisfying conclusions is *not* a minority preference, thumped into us by the mallet of Western culture. If so, why might human beings tend to like the well-defined ending best? After all, the twentieth-century fad for inconclusive narrative is actually more like life, which is always going on...and on...and on. If we like art that represents the world, you'd think we'd like poems or stories that wander off.

However, out of the capacity for movement, we developed the capacity for purpose. As hominids, we evolved increasing abilities to form longer-range intentions involving others, and to manipulate the world's materials to our purposes. We have feedback loops in our brains that cause us to stamp something as "done." This satisfaction rewards sustained effort. One of the most difficult malfunctions that a person can live with is the one that disrupts these loops and creates obsessive-compulsive disorder—the feeling that an action has not been completed, no matter how often it has been repeated. It is a terrible state of constant anxiety to live in.

So it is not surprising that we like works of art that leave us feeling the end has ultimately been reached. This feedback loop is perhaps the e-motion that underlies Kant's view of art as purposeful without purpose, and finality without end.

TO STAND in an aspen colony is a mildly hallucinatory experience. The trunks, so straight and tall and uniform, glow faintly green—aspen can process sunlight with chloroplasts in its bark just as it does with its leaves. When you drive along highways in my region, the arrays of slender trunks seem to form edges, bars, diagonals leading to the interior—an arbitrary grid imposed by the brain. But in the middle of the stand, that grid is lost. There are no straight lines. If you turn slowly, you seem to see curves and arcs that suggest the lines drawn by a force field. It's like being inside a mandala.

Aspens may be not only the largest organisms on Earth but also among the oldest. The tree is a colonizing species, the first to push into bare, moist, mineral soil left as the glaciers retreated. There is little of that soil available now to nourish seedlings, so aspens in Alberta today rarely reproduce from seeds. They rely on their network of roots to poke and propagate. Some of the aspen stands here could be 10,000 years old.[21]

The word "aspen" is also very old. Linguists have traced it back to Proto-Indo-European, the language that gave birth to daughter languages from Sanskrit to Greek to English. The vocabulary deduced for this ancient language paints a picture of the region its first speakers lived—a people who needed to speak of wolves and ice, of barley, goats and udders, of beech trees and aspen.[22] It was a landscape very like the aspen parkland I now call home, here at the edge between prairie and the vast, circumpolar, boreal forest. Like humans, aspens have girdled the globe.

AT THE END, when Alzheimer's had taken virtually all narrative from my father, he no longer had any sense of where he was, no map, no home. He walked, walked, walked, seeking that elusive sense of completion that says you have arrived. His memories were a jumbled custard, words a clotted tangle. The only sentences he could speak fluidly were his poems—the words that had become so much a part of his procedural memory that he could assemble the complex flow of movements in the tongue's many muscles, one phrase following the other in its chiming sequence. If I gave him a phrase or two, he would happily recite whole chunks of his long rhymed narratives. I would recite along, nudging him occasionally. At the end, I knew his poems by heart myself. He was like an elderly pelican that had passed along the map home to someone else who could follow it for him.

Poems remember for us where we have been.

Politics as Phase Space

TEN

Elections come and go, and I live in
a province that stays the same shade
of blue, the historic colour of the
Conservative Party, a party traditionally associated more with the interests of
business than artists. Alberta can look
remarkably monolithic from a distance—
a place of ornery uniformity. Under this
sky-blue regime, what is the proper role
of an artist? What kind of space do we
(should we) create and occupy? I always
struggle with this question, not being
temperamentally inclined to rebellion but

afflicted by persistent guilt. If I don't care for the mindset of
a governing party, shouldn't I do something about it? If so,
what? And how?

Throughout the twentieth century, we have tended to view
the relationship between artist and state along a single axis. Its
poles are, at one end, the sanctioned art of a totalitarian state
and, at the other, the dissident—the artist who is prepared
to risk liberty and more to oppose the government by the
means of art. Maundering off to one side, but irrelevant to
the discussion, are artists who have turned their back on the
situation and say that art and politics inhabit different worlds.

This axis is a legacy of the great political debate of the
last couple of centuries, the debate between democracy and
totalitarianism. But we know it's a simplistic one. I've come
to think that a better picture of the artist-state relationship
might be what mathematicians call a "phase space." A phase
space is a way of turning numbers into pictures, of taking
"every bit of information from a system of moving parts...
and making a flexible road map to all its possibilities."[1]
For a relatively simple phase space picture, think of the
butterfly shape that you see in textbooks on chaos. That's
a two-dimensional phase space, and it's made by tracking
only two variables: location and time. But there can be phase
spaces with three, four, or twenty dimensions, depending on
the number of variables you want to track. (It's impossible
to actually picture a phase space with more than three
dimensions, but mathematicians can manipulate them.)

A phase space, says mathematician Ian Stewart, has one
important benefit for mathematicians: "Their main role
is metaphorical. They embed what actually happens in a

structured realm of what might have happened instead."[2] That butterfly pattern shows you where you are likely to find the system at some point in its progress—and where you aren't likely to see it, ever.

How do "phase space" and "variables" translate to the relationship between art and politics? Here on my desk I have two books by poets who were intimately involved with the politics of their time. One is Pablo Neruda, the great Chilean poet whose passion for social justice led him to speak out against repressive regimes in his poetry and also to take an active part in politics—even to the point of running for president.

The other figure is less obvious, perhaps more problematic, as a political artist. Six hundred years ago, Geoffrey Chaucer had an intimate, practical, almost cozy relationship with power.[3] Brother-in-law to the powerful John of Gaunt. Busy civil servant, administering customs offices for the Crown. Beloved court poet, entertainer to the Queen and her ladies. His great work, *The Canterbury Tales*, is brilliantly innovative—a tale of tale-telling, set in an England that seems bawdy, hierarchical and essentially sunny. You could easily read it and never realize that the country he was living in was on the edge of civil war, that cruelty was arbitrary and poverty rampant. How unlike Neruda's direct condemnation of governments and armies.

In terms of words like "compromise" or "dissidence" these two great poets would seem to be at opposite ends of the spectrum. But a better analysis is that they inhabited different spaces that weren't defined simply by a personal willingness to compromise. And I can think of four significant variables that define the artist's phase spaces in any region of space-time— including the Alberta of today.

First there is the basic stability and justice of the political system. How bad is it? For ordinary people living in Neruda's Chile and Chaucer's "Engelonde," it was pretty bad. In Alberta, the economic situation is very difficult for a marginalized portion of the population, but there's a larger number of ordinary people for whom things are fine. They can negotiate their day-to-day lives with a reasonable expectation that they will know what those lives will bring them.

The second variable is what political options are available. Chaucer had no alternative structures to point to—no democracies, no socialist states, none of the more modern experiments that Neruda could point to as an alternative to a repressive regime. In such a situation, what can you do but say, "Well, let's make this system work as well as possible." Oddly, I feel rather like that about Alberta. Cross as I can get at politics here, I can't see a significantly different political system that I'd put in place. Is this simply lack of personal imagination or a real limit we are up against?

Thirdly, we have the perceived role of art. The governments faced by Neruda saw art as dangerous to them. The kings and courtiers served by Chaucer didn't pay him that compliment, but they took his work seriously. To be cultured was a good thing—an entertainment in the original French sense of *entretenir* (to hold something in common, to share). In Alberta, artists are not important enough to be dangerous; if we are to be entertaining, it is in the debased sense of pastime, amusement. In the eyes of the government that has prevailed for the last decade or two, art is just one of the range of options that citizens can select from to pass their time—no more or less valuable than a night at the casino.

The fourth variable is the individual artist's temperament. What would a Neruda have written if dropped into Chaucer's England? What would I write if dropped into Stalinist Soviet Russia? I have an uneasy feeling I might have settled for producing acceptable poems in exchange for a dacha on the Black Sea.

Given these variables and how they define Alberta's phase space, what seems likely or unlikely here? Well, for one thing, it seems virtually impossible that a piece of art—a poem, a painting, a play—will galvanize political change. It

is all too possible that we will write only personal lyrics—
lovely, intelligent, experimental ones, no doubt—and paint
landscapes and abstracts of the same description. That
our playwrights and theatres will attract audiences only to
comedies, because they are entertaining.

Is this so bad? Is it good? Or just what we have to live with?
What does an artist do, if you don't want to retreat into the
position that art and politics are different realms? Or if you
don't have the rebel's temperament?

For one thing, I think you look at what artists make their
art about. Artists always have new content. It's around us
every moment of every day, and it is inherently political. Look
at Chaucer again. *The Canterbury Tales* may not look political
from here, six centuries further on, but that is only because we
have narrowed and redefined what politics covers. In fact, the
content of Chaucer's great work is about the wide gap between
how things are and how they should be. The hierarchy doesn't
work as it should, he says, particularly with regard to the
Church. In fact, it's a wonder to me that he wasn't in serious
trouble for his portraits of those rascally monks and friars—
and in fact, he might have been.[4]

Formally innovative as Chaucer was—and no one had seen
anything like *The Canterbury Tales* in English poetry before—
his greatest innovation was content. The characters and lives of
contemporary people made his work revolutionary.

If I was a painter I wouldn't want to paint abstracts and
landscapes. I'd want to paint the teenage mothers I see on the
Westmount bus with their loved but too-fat babies and their
bags stuffed with school books and diapers. That's politics. If
I was a landscape artist I'd be painting scenes from a clear-cut
forest, or the lines of a strip mine.

Of course, content in and of itself will not create political change. How many people in today's Alberta read a poem or look at a painting? Too few to start a revolution. I want to accept that but not accept that it means there is no useful or necessary connection between art and politics. The right images could help catalyze social change.

Which brings me to the second response that artists can make in our particular phase space. I think we do have to become directly involved in the political process—not to lobby for increased arts funding because, frankly, that won't do much good. But to change the people who lead, which we can do (as Chaucer's contemporaries could not). We may not want to create a new world order; we may even accept that governments have relatively little room to manoeuvre when it comes to balancing budgets and delivering programs. But governments do have immense room to manoeuvre when it comes to the spirit in which they lead us.

When it comes to the third variable—perceived importance of the arts—we could live in a province where political leaders feel the creation of music or artwork is something to be actively fostered, that plays are something to go to voluntarily, that citizens as a whole are better off if there is lively production of artwork here. In short, leaders who personally feel that art isn't a merely a potential consumable but what we share, that which holds us together.

The Canadian population will not rise up en masse demanding better funding for the arts. Politicians know this. But if we have politicians who personally believe in the importance of soul, art and creative energy, how different our phase space would feel. How much would seem possible then! So finally, I come back to that last variable, the individual temperament of the artist. And to Neruda, willing to take a seat in the Senate of Chile and run for president. Who of us is willing to do as much here? What might happen?

ELEVEN

Brain Surgery

The story goes that Margaret Atwood was buttonholed at some social function by a brain surgeon. He told her he was interested in writing, too, and intended to take it up after he retired. She snapped back, "And when *I* retire, I'm going to take up brain surgery."

Writers love this story. Perhaps it warms our hearts because half the people we meet, from brain surgeons to cab drivers, tell us how they want to write a book—as if it was roughly equivalent to wanting to tour Spain this year. All they need is a little holiday time. Or perhaps we love it because we are secretly insecure. Is it really so hard to write a book? I write poetry because I find it easier than the options. It *would* be much harder for me to be a brain surgeon or a cab driver.

One of my other very favourite stories concerns William Hamilton, the great mathematician who gave his name to the Hamiltonian equations that describe the total energy of a system. Hamilton wanted very much to write poetry and it took his friend William Wordsworth to point out tactfully that his talents did not lie in that direction. "You send me showers of verses which I receive with much pleasure...yet we have fears that this employment may seduce you from the path of science."[1] Hamilton was a phenomenally creative, inventive thinker, and part of me thinks that anyone who could handle the symbolic language of equations so jauntily could surely crank out an equally good poem. That he found it difficult is a relief. It seems to confirm that I'm doing something that does indeed require something special.

Yet, I also argue with people who tell me, "Oh, you were *born* that way," as if I had won some odd lottery. No one says that to the brain surgeon. People assume he may have arrived on the planet with some useful predispositions that were subsequently developed, but those predispositions might equally well be applied in engineering or a biochemistry lab rather than to the manipulation of scalpels. The decision to go down one avenue rather than another is attributed to relatively prosaic forces of inclination; brain surgeons do not assume their choice has to come from a celestially dictated organization of their brains.

So is the ability to write poetry innate or acquired? If it's the former, what kind of special organization might be required? If it's the latter, how do we attend to the schooling of poets?

IF POETRY IS INNATE, you'd expect to see a wide divergence of talent emerging at a very early age. Surely infant prodigies clearly illustrate a divine tap on the head. However, in comparison with music and mathematics, there are remarkably few prodigy poets. Mozart was remarkable for being able to compose minuets at the age of five. Pablo Picasso at the age of thirteen was convincing a jury to allow him to attend Barcelona's School of Fine Arts. Poets are laggards by comparison. My own first lines of verse, composed at the age of seven ran like this:

> *a furry coat has the bear to wear*
> *the tortoise a coat of mail*
> *the yak has more than his share of hare*[2]
> *But the pig has a curly tail*

My parents hailed this work as a great achievement, but even they wouldn't have considered putting me on a stage to recite it. I hasten to add that I'm *not* setting myself up as the literary equivalent of Mozart. But a genuinely great poet, Robert Frost, at the age of sixteen, was producing such gems as:

The 'tzin quick springeth to his side.
His mace he hurls on high.
It crasheth through the Spanish steel
and Leon prone doth lie.[3]

Such a stanza hardly suggests a future laureate. At the same age, Mozart had written a very creditable and popular opera, *Lucio Silla*, that is still being performed.

The trend does not apply only to star artists; it is fairly general. When you walk through the halls of any arts-focused high school, you will see visual art on the walls that is sophisticated and technically accomplished; from the music studios, you will hear music played at a high level. Young people can be wonderful actors. Their performance in all these artistic disciplines will be consistently more impressive than their poetry efforts at the same age. So if poetic ability is innate, it requires an unusually long gestation period.

PERHAPS this kind of delay indicates that there is a lot of practice involved in mastering poetry. Research indicates that the mastery of any skill takes around 10,000 hours of practice. Put in enough time, and you, too, will become a genius. In fact, this may be part of the cognitive processing that underlies the remarkable capacities of savant syndrome—people who have trouble with many ordinary activities but have "islands of genius." Some savants can play a complex piece of music after hearing it only once; others might be able to calculate the value of pi to record-setting decimal places. Savants seem to like doing something again, again and again. And because they don't do other mundane things with their brains, they spend phenomenally large amounts of time on the tasks that interest them.

It's fairly clear what a musician can spend 10,000 hours on: practicing scales, rehearsing pieces over and over, learning the theory. But what does a poet spend her 10,000 hours on?

In fact, we put in our most important apprenticeship very early, before the age of six. As functional speakers of language, we were learning the rhythms and sounds of speech in our high chairs. Our ability to co-ordinate the muscles of the tongue and lips, the rush of air from the lungs, the complexities of syntax and the social rhythm of speaking makes six-year-old Mozart at the harpsichord seem almost ordinary in comparison. However, this is a mastery that poets share with just about everyone else on the planet; in matters of language, virtually all human beings are prodigies.

However, although poets learn this procedural basis of their art very early, we do relatively little with it for a long time. Then poets can effloresce rather suddenly. The first surviving poem by John Keats was written at the age of nineteen; within seven years he was dead, having produced poems that were to become some of the most influential in English literature.

I suspect it is not a coincidence that effective poetry doesn't emerge until around the same time that development of the brain's frontal lobes is complete, which happens in our early twenties. During puberty, a wave of development in the frontal lobes adds massive numbers of additional neurons. Over the following decade or so, there is a steady increase in the myelination of this region. Myelin is the white layer that sheathes the axons of neurons, insulating them and allowing them to conduct impulses more efficiently. The frontal lobe's executive functions (such as planning, controlling behaviour, and organizing multiple tasks) aren't working at peak until the process of myelination is finished. It may be that poets require this type of processing more than other young artists do.

Savant syndrome may provide additional support for the importance of general brain connectivity to poetry. Savantism seems to be associated with an imbalance between the left and right hemispheres of the brain; experiments can make temporary savants of normal individuals by using transcranial magnets to handicap the left hemisphere.[4] In such experimental situations, people can spontaneously draw remarkably realistic scenes—an ability that normally has to be taught. It is actually a little surprising we *can't* draw like this more easily, since our brains have all the information necessary to do so. However, we can't seem to access it. The left hemisphere gets in the way, imposing patterns of meaning on data, filtering out the raw information. Unlike artistic savants, most of us tend to be more aware of the meaningful whole than its constituent parts.

Savants' remarkable abilities typically bloom out of nowhere, unbidden and apparently untrained, between the ages of five and eight. There's considerable crossover between the areas in which infant prodigies and savants excel: music, realistic drawing, mathematical calculation and mechanical/spatial skills. But poetic ability in savants is very rare, just as it is in child prodigies. When it comes to writing poetry, the capacity to work with a "meaningful whole" may be essential—which in turn requires maximal connectivity throughout the brain. Whether it is the more efficient connectivity of the frontal lobes in particular, or more generally throughout the cortex, it seems that something about being able to draw on the whole cortex—left and right, back and front—is essential to writing good poetry.

This is not simply the same as needing more life experience in order to write. We often say that young writers just haven't gone through enough to have much to say. But what we really need as writers is not so much life experience as certain capacities to think about it in complex ways. This capacity is somehow bound up with the connectivity of language. For even if language is not (as extreme postmodernists imagined) *all* of thought, neither is it a tidy little module tucked into the tail of your left temporal lobe. It's a faculty that is massively connected to other regions of the brain and draws on all of them.

WHEN I WAS IN HIGH SCHOOL, writing secret poems that tried to describe adolescent angst, I helped a friend with a high-school essay. He had chosen one of Leonard Cohen's poems, "A Kite is a Victim," for his topic. During our animated discussion it suddenly struck me that, along with all the images inside the poem itself (kite as poem, as fish, as "contract of glory"), the kite could also represent something like soul, something connected to the physical world and yet beyond it. This concept was not specifically mentioned by the words of the poem, which are more obviously about the kite as a meta-phor for poetry, so it struck me with a metaphorical blow that concepts can overlap and resonate beyond themselves. It was a eureka moment that still rings in my ears. Suddenly I realized there was another level to writing—that beautiful combina-tions of words could point to something beyond themselves.

Until then, I had been good at English in the usual way. I could read easily; I could understand and use metaphors or puns or other figures of speech. But there was a literal quality to my most fanciful writing that reflected how I read literature. As a little girl in grade four, I had loved the Narnia stories. But in all my reading and rereadings I'd never noticed their Christian iconography, even though I was exposed to exactly the same imagery every week in Sunday school. At twenty-one, I reread *The Lion, The Witch and The Wardrobe* and wondered how on earth I could have missed it.

IMAGINE we were able to engineer a poet's brain. What would you do? Many people would assume it must be quite different from the brain surgeon's or any other scientist's.

High-school chemistry class. We are partnered off at lab benches with Bunsen burners and small sinks, and handed test tubes and a small lump of blue crystal. First, we are to heat it in the flame of the burner, watching it turn to a greyish powder. Then we are to add water to the test tube, whereupon it turns to a bright blue liquid. I stir the vivid liquor with the end of my pen and write down the observations as the teacher has told us to—what did we see? what happened at each stage? I absent-mindedly stick the end of my pen in my mouth and wrinkle my nose at the bitter taste. When the teacher asks us what we've observed, the other lab pairs answer with the same routine things—changed colours, consistencies, quanti-ties. I am pleased to hold up my hand and announce something different: "It tastes like copper pennies!"

The teacher leaps up in horror. "You don't mean you *drank* it," he gargles and runs down the aisle, prepared to rush me to the nearest stomach pump. I barely have time to point out that the tube is still full, and all I did was get a little on my tongue. Relieved, he uses this as a teaching opportunity to point out that, in future experiments, we should depend on all of our senses for observation *except* for that of taste. I will go through my life remembering the particular qualities of copper sulphate.

The point of our high-school experiments was to train us in laboratory techniques. The teacher didn't expect anything novel from them beyond the idiotic unpredictability of the students themselves. We were being taught to observe, to *experience* (which is related to the word "experiment") something for ourselves—useful habits for any writer.

Two stereotypes underlie the idea that the cognitive processes used by scientists and poets are fundamentally different. One is expounded by Robert Graves, that experimental research involves "a series of routine experiments in the properties (say) of some obscure metallic compound." [5] Certainly our experiment in the high-school lab was routine, but no more so than the sixteen-year-old Robert Frost's experiment with predictable meter. But scientists don't do experiments where they *know* the outcomes; they are looking for surprises. As high-school students we are surprised by the experiments we haven't done ourselves, but once we know that copper sulphate goes bright blue, we don't need do it again and again.

Which leads to the second, and contradictory, stereotype— that science is relentlessly about novelty, inventing something that has never been seen before. Science leads to new things, of course, but in the same combinatorial way that writing leads to new poems. You don't have to invent a new language for either endeavour. You build on what has gone before.

If you're building a poetry brain, you need a fundamental ability that is also essential in science—the ability to simultaneously maintain conviction and doubt. Only then can you overcome the difficulty described by Francis Bacon of making progress while nonetheless retaining the skepticism necessary to ensure the correctness of your results.[6] "How can you take an idea seriously enough to delve into its consequences while nonetheless suspecting that it might be incorrect?" It's the same ability a poet needs to take what she is doing seriously but be prepared to chuck it overboard if something else proves more effective.

IF POET'S BRAINS aren't so very different from scientists', are they distinguishable in some other way? One romantic notion is that the brains of poets (and other artists) are so profoundly different from the rest of the lucky "normals" that they qualify as "mad." It's a viewpoint that irritates me. I feel condemned to minor poetship because I don't have some flamboyant addiction, while I know intelligent and creative people whose work has been derailed by dealing with mental illness.

We're in the process of learning that conditions like schizophrenia, bipolar disease, obsessive-compulsive disorder and their cousins are not so much things that are "wrong" with the brain but extreme ends of conditions we all inhabit. The astonishingly uniform prevalence of schizophrenia across cultures, for instance, indicates that it involves something that can rather easily malfunction.[7] In schizophrenia, input from the senses seems to get overloaded and the brain's circuits create internal representations that are so strong, so compelling, they seem real. Somewhere in our recent evolutionary history, humans developed the ability to make mental representations of things that weren't there, as part of the overall development of a complex capacity for memory. Creators of all kinds, including poets, depend on this faculty; without it, you could not have the misfiring typical of schizophrenia.

Similarly, obsessive-compulsive disorder is a normal state for new parents, who check and check and check to make sure their infant is safe. But when someone's brain goes into the state where that switch can't be turned off, it can become a lifelong, agonizing condition. Once again, a mild case is useful to poets; it keeps me reading and rereading lines to be sure they are right.

It's not that some people are mad and some are normal. It's that we all deal with the same brain systems that can function all the way to the edges of malfunction.

SO DOES *ANYTHING* "define" a brain organized for poetry?

Here I am on a Monday morning, wondering what this might be. Unfortunately, the only brain I have experimental access to is my own. It's not a large enough sample, but it's all I have handy. So what might predispose me toward this particular activity?

I know that it's not that I'm more observant than other people. I don't perform particularly well on those tests where you have to pick out the differences between pictures. I don't *feel* more than other people do, either. I have a rather equable disposition on the whole, and on the Myers-Briggs tests, I come down on the side of "thinking" as opposed to "feeling" in my decision-making processes.

Perhaps the difference lies in my compulsion to translate observations or feelings into language. Maybe these pathways in my brain are worn into deeper grooves. But that wouldn't necessarily make me more inclined to become a poet than, say, a novelist or an historian or a journalist.

I know that part of why I like writing poetry rather than, say plays, has to do with the fact that I'm an introvert. It's not that I don't like people—you are lovely. But I find you very tiring and hard to tune out; I feel I have to pay attention to each one of you and make you all comfortable. It's a relief when you are not in the room. A collaborative working place like a theatre or classroom exhausts me. This particular quirk may come from having a slightly overactive amygdala— something that has been observed in the brains of shy children. (The amygdala is the part of the limbic system that stamps experiences with an emotional tag, especially those related to fear.) Or it might have something to do with the process of habituation, the capacity of brains to rule out continuous stimuli (like the rattle of traffic outside) so you can focus what's changing and needs attention. Research indicates that creative people tend to take much longer to habituate to a stimulus—a characteristic they seem to share with people who have schizophrenia.[8] Whatever the reason, I am content spending an amount of time by myself that would send other people screaming into the street for company.

The idea of a hermit-poet brain would be slightly intriguing, because there is suggestive evidence that language skills and sociability are normally tightly coupled. Williams syndrome, for instance, results from a genetic mutation that results in significant impairments to spatial and other mental abilities. However, children with the syndrome have excellent language abilities and are exuberantly social. Autism spectrum disorder tends to reverse this, causing both language and sociability to suffer. So poets would have to mildly dissociate the two capacities that are normally linked by having strong language ability but a lower desire for social life.

There *may* be a slightly higher-than-average tendency for poets to be introverts. After all, you have to be sufficiently willing to sit by yourself for long enough to assemble words. But once again, introversion is neither a sufficient nor necessary component for a poetic brain. Lots of introverts are *not* poets; many poets are gregarious.

If we are looking for a *sine qua non*, it clearly has to do with language. I think the only aspect of language that may distinguish poets from non-poets is a predilection for playing with the sounds of language long after most children say, "Okay, got *that* figured out." It's a kind of slightly arrested development that keeps us hyper-conscious of how words sound and connect.

SO IF A POET'S BRAIN isn't so very distinctive to start with, what shaping do they need to create good poetry? What should we spend our 10,000 hours on? It's the question that schools of creative writing must try to answer.

When I was going to college, there were no creative writing degree programs in Canada that I knew of. My college did offer one creative writing course in fourth year. I had been dying to take it but chickened out at the last minute because the professor terrified me. Today, creative writing programs have sprung up across the land like a wave of neuronal development, and the debate over whether they are good or bad for writers springs even more vigorously. (Eudora Welty, when asked whether creative writing programs squelched writers, responded tartly that in her opinion they didn't squelch nearly enough of them.)

However, the professional training of poets is not just a modern idea born of an insatiable need to hang a certificate on every wall. The Bardic Colleges of Ireland flourished for centuries, turning out workaday poets, not necessarily inspired ones, who found a niche with each royal family to celebrate their deeds and genealogies. Bards were steeped in history and tradition, as well as technical requirements and poetic conventions. The bard was, in fact "a professor of literature and a man of letters, highly trained in the use of a polished literary medium, belonging to a hereditary caste in an aristocratic society, holding an official position therein by virtue of his training, his learning, his knowledge of the history and traditions of his country and his clan....He was often a public official, a chronicler, a political essayist, a keen and satirical observer of his fellow-countrymen."[9] This role of the trained, professional, official poet may actually go back to the days of Indo-European tribes from whom Celtic, Greek, Indian and other cultures have descended.[10]

In these degenerate days, the potential for becoming a "professional" poet is much more limited. You'll wait a long time with your creative writing degree before seeing an ad that reads: "Poet wanted. Competitive salary and benefits." Few clans or corporations are willing to pay for a professional poet.

(And certainly very few clients are willing to risk the boils that, according to legend, could be raised by a particularly scathing satire from a dissatisfied bard.) The jobs available to you will seldom be to *write* poetry; instead you will mostly earn money teaching others to write it, and publishing your own work will be merely a confirmation that you have the street cred to teach.

However, assuming that you will graduate and go off to write poetry for the usual starvation wages outside the academy, the question is—did the time spent in the training program help? The main objection to such programs is that they encourage a kind of group-think, that the poems produced by their graduates have a certain faddish sameness. However, human culture has a certain inherent faddishness in all generations and cultures; the minor poets of Elizabethan England tend to sound much alike, and we can't blame *that* on creative writing courses. The good thing about courses is that otherwise isolated people can get together to find others who will actually read their work; they gain the energy of community, learn basic techniques and meet people who will be useful in their later lives.

However, creative writing programs are neither necessary nor sufficient in the way that professional courses are for a brain surgeon. Surgeons need to demonstrate that they have mastered a body of knowledge, and to do so they need to be in a classroom organized to deliver, test and accredit. Poets, on the other hand, can come from any igloo or hermitage, provided they have the opportunity to absorb sufficient amounts of poetry. And this doesn't mean a poet needs to read every classic, every "great" in his culture. He needs a deep familiarity with a small number of poems first. The process is a bit like learning language as a whole—we infer rules of speech from a limited, finite sample of sentences we hear before we are two. We don't need (and couldn't possibly listen to) all possible combinations of words in our language before becoming competent practitioners.

Of course, wider familiarity with literature gives context and inspiration and an impetus for excellence. Poets often read other poets and think, "Wow! I could take that (*The Odyssey*, Spanish *glosas*, the word "infinity") and apply it to *this*." Brain surgeons really shouldn't try this kind of process on operating tables. If you are going to become truly and consistently excellent, you will go on learning as much as you can about the vast realms. But there is always the possibility that someone with relatively little of this kind of training can toss off a one-and-only poem that reaches many people—poems like "High Flight" in our school readers. Brain surgeons, by contrast, cannot hope to perform one magically lucky operation before they've finished the course.

However, reading other poets isn't quite enough. The inability to practice poetry as a paying profession is actually our secret weapon—we are forced out into the world to do the kinds of things that other people do. Chaucer the civil servant. Keats the doctor. Alice the office worker. I don't argue that poets must do exotic things or travel to exotic places to be interesting. But we need a kind of interpenetrability with the larger world to write works that will engage with an audience. We cannot simply write about writing.

This is why I personally feel more of a kinship with the self-taught poet. Of course, I may simply be afflicted with the residue of a romantic view of the divinely inspired poet—an

idea I tend to reject at a conscious level. Or maybe I'm just
envious of networks that I don't feel I belong to. But in a
world that has become vastly more segmented, the danger is
that creative writing graduates do not go out to serve a wider
public. Their world, their world view, their connections and
plaudits come from the same place they learned their trade.
Creative writing programs don't even teach nearly enough
of one skill the Bardic Colleges would have considered necessary—i.e., performing for an audience.

Poets are like scientists who have to connect their work
with experimental success outside the academy, or carpenters
who have received their journeyman's tickets and go out to fit
cabinets in a real kitchen.

WHEN ALL IS SAID and wondered, however, I know that what
I do *is* different from what the brain surgeon does and—*pace*
Margaret Atwood—it's easier. Poetry does not depend on the
huge edifice of innovation and knowledge that the scientist's
vocation does. What I do could have been done just as well
around the campfire of the OMO people, a hundred thousand
years ago. And it could be done well by many, many people
today—in any class where I give a poetry workshop, there will
be at least one student with a "gift," the slightly tweaked sensi-
tivity to language that could be the foundation of poetic ability.
That brain surgeon just *might* retire and write a good book.
This does not reduce the value of the ability to write.

We do not value what comes easily to human beings,
because we do not realize how enormously complex it is to
move, to recognize a pattern, to tell a story, to love. In the last
century, we felt that we had made poetry "better" by making it

hard. The greatest gift of science in our time may be to help us understand how wonderfully sophisticated our simple abilities are, how long it took to evolve them, how they fit into the natural world as part of a continuum.

As scientists continue their investigations we will come to understand in greater and greater detail how the brain does its easy things. But science is also increasingly aware of its limitations—that the sheer volume of information required to understand even a single experience exceeds the descriptive capacities of laws and equations. "Events are denser than any possible scientific description," writes Nobel-winning neuroscientist Gerald Edelman.[11] Ultimately, the poet's description is as complete a description of experience as the scientist's.

FROM THE BRAIN inside this skull, looking out, I make observations of my world. I move words into arbitrary combinations, testing them for sound, testing them for a connection to my observations. I let their connections echo through the net of memory I have spun over decades of learning. I measure and fit, listening, listening all the while to these new artifacts from old words, hunting through combinations of the familiar for the ones that have not been made before.

A Poem
is How We
Felt When...

One last story. A steeply raked lecture
room in the engineering building at the
University of Alberta. At the bottom of
the tiers of seats, like a ball at the bottom
of a pinball machine, is a poet. The celebrated Canadian philosopher-poet Jan
Zwicky has come to talk about the relationship between poetry and science, and
she speaks eloquently about the need
to incorporate the emotion that drives
lyric poetry—an intense attention to the
natural world—into the rationality of
technology, injecting this emotion into
decisions that will affect Earth.

Her talk is inspiring, scholarly, passionate. This is a woman who knows her Wittgenstein, not to mention her Hegel and Habermas, her Descartes and Kant. I would not want to argue points with her—I'm way out of my depth. And who wants to argue, anyway, that lyric observation and love of Earth should *not* be incorporated into engineering choices?

Yet I leave the hall feeling vaguely argumentative. I finally put my finger on it hours later. Her thesis was built on the old assumption that art is emotional and science (along with its step-daughters technology and business) is not. But anyone who has ever worked in a lab or been through something as dull-sounding as the regulation of the utility business knows that all these human activities seethe with emotion. And whenever I sit with some young friends who are early in their science careers, I feel their excitement and enthusiasm to be as contagious as any artist's.

"Don't call *me* unemotional," says the scientist to the poet, looking around for something to hit her with.

The human world is awash with emotion. Poets merely give that feeling a little structure.

IT IS MORE THAN THREE YEARS since Barack Obama's inauguration as president of the United States. In the tangled pulpy mass of economic meltdown, we have forgotten the meteoric optimism. Opinion polls buck leaders off like the pinball machine's inevitable rattle downward. We are caught in the interminable flux, the constantly curdling custard. My Canadian city seems fixated once again on how we can't get our side streets ploughed clear of snow.

What can a poet do? We cannot stop the flux. We can only reach into it, pull out a moment and all its feelings, organize it using all the "rational" tools of language and pattern. Then we send it back into the flux as a small machine, another tool for thinking with. A poem is how we felt when...

THE PLAY IS OPENING
Morning of US Election Day, 2008

The restless news on battlements.
Kings go to coronation or to slaughter.
 November night
 hunches its shoulder, a hump
 of silent plumage.

Soldiers move in deserts. Deployments
and daggers, ghosts before dawn.
 The old moon startles. Magpie
 crescent, slash carved on black.
 A planet casts a bright eye.

Campaigns. Days of cast iron, cast ballots,
scrying glass and scratched bones.
 The bird of dawning
 broods on the nest of the east
 pale breast, slate-blue back.

Oneiromancy, the cock crow of news channels.
Our dreams stumble into wakefulness.
 Rose-barred mottlings of cloud,
 hawk hoverer, wing tips flared
 from horizon to horizon.

Rumour, full of tongues, rattles.
Will a black man rule or fall?
 A wedge of red deepens
 between rooftops, childhood's robin,
 winter's scarlet waistcoat.

The birds of dawning fly around the world
fiery migrants. We rise
 to clouds a dun custard,
 the daily gray of pigeons,
 and hope.

Notes

The Magpie's Eye

1. The quotes from Ted Hughes and Robert Graves can be found in Geddes, 561 and 581.
2. Johnson, 220.

ONE That Frost Feeling

1. Frost, 701.
2. Emotional tears include proteins and hormones not found in reflex tears, and also much higher concentrations of manganese. Since manganese concentrations have been found in the brains of people with depression, some theorists feel that the lacrimal gland (which produces tears) may help remove this build-up. See Lutz, 101.
3. See Barnes and Thagard, "Empathy and Analogy."
4. The mechanisms of empathy are outlined by Jean Decety and Claus Lamm in their paper, "Human Empathy through the Lens of Social Neuroscience." See also Sandra and Matthew Blakeslee's *The Body has a Mind of its Own*, Chapter 9, for a discussion of the role of mirror neurons in creating the empathetic response.
5. See Decety and Lamm, 1150.
6. This is a major theme of Ellen Dissanayake's book, *Art and Intimacy*.
7. William Wordsworth articulated this definition in the preface to his *Lyrical Ballads*.

TWO   Metaphor at Play

1.   Turner, Chapter 2.

2.   See Hofstadter, "Analogy as the Core of Cognition."

3.   Anthimeria is the substitution of one part of speech for another (such as a noun used as a verb). Zeugma is a figure of speech in which a word stands in the same grammatical relation to two other words but has a different meaning with respect to each of them. For example, Alexander Pope's image of someone who is prepared to "...lose her Heart, or Necklace, at a Ball."

4.   See Elena Semino and Gerald Steen, "Metaphor in Literature," in Gibbs, ed., 242.

5.   Dobyns, 17–18.

6.   Walcott, 5.

7.   McAdam, "Memory," in *Cartography*.

8.   Dante Alighieri, Canto XVIII, Line 103, 296.

9.   See Zoltán Kövecses, "Metaphor and Emotion," in Gibbs, ed., 392.

10.   See Ning Yu, "Metaphor from Body and Culture," in Gibbs, ed., 257.

11.   Damasio, *The Feeling of What Happens*, 136.

12.   Seana Coulson's paper "Metaphor Comprehension and the Brain," discusses the history of research on right brain/left brain comprehension of metaphor, in Gibbs, ed., 177–91.

13.   See Lynne Cameron, "Metaphor and Talk," in Gibbs, ed., 200.

14.   For a discussion of the "career of metaphor" concept, see Dedre Gentner and Brian Bowdle, "Metaphor as Structure-mapping," in Gibbs, ed., 116–19.

15.   See Alan Cienki and Cornelia Muller, "Metaphor, Gesture and Thought," in Gibbs, ed., 486.

16.   See Papagno et al., 78–86.

17.   Holub, "Not so Brief Reflection on the Edict," in *The Rampage*, 34.

18.   Dobyns, 21.

19.   See Budiansky, 53.

20.   See Penrose, *The Road to Reality*, 73.

21.   The study of how people process metaphors and similes is outlined in Sam Glucksberg, "How Metaphors Create Categories—Quickly," in Gibbs, ed., 76–79.

22.   See Gibbs, ed., 121.

23. See Page, 42–43.

24. Roald Hoffmann points out the different metaphors used for science and art in *The Same and Not the Same*, 87ff.

25. Koch's book, *Wishes, Lies and Dreams*, is a great resource for teaching children how to write poetry creatively rather than relying on stereotypical exercises.

THREE The Holographic World

1. The double-slit experiment is a classic puzzler first carried out by Thomas Young in the early 1800s. The experiment (which has been repeated many times since with different kinds of particles) shows that light behaves differently if it goes through a single slit in a barrier to a screen behind than if it goes through two side-by-side slits. It's as though an individual photon "knows" what set-up it's dealing with and behaves in an experimental set-up with two slits as if it were part of a wave, and in a set-up with one slit as though it were a small discrete particle. The details are well described in Feynman, Chapter 3.

2. Quoted in Capra, 317.

3. See Bekenstein, 61.

4. The study by Reuven Tsur is outlined in *Toward a Theory of Cognitive Poetics*, 217.

5. MacEwen, 130.

6. The phrase comes from Brian Greene: "In a universe governed by string theory, the conventional notion that we can always dissect nature on ever smaller distances...is not true. There is a limit and it comes into play before we encounter the devastating quantum foam." Greene, *The Elegant Universe*, 156.

7. See Aczel, 198–202.

8. Whitman, 72.

9. Quoted in Capra, 317.

10. See Feynman, 128.

11. See Randall, 224.

12. See Smolin, 104.

13. Ibid., 177.

14. See Dissanayake, 212.

FOUR Points on the Line

1. Aristotle's concept of the golden mean is outlined in Book 2 of his *Nichomachean Ethics*.

2. Robert Jourdain discusses the brain pathways used in processing sound in *Music, the Brain, and Ecstasy*, 28.

3. The sum-over-histories approach is explained in Richard Feynman's book, QED: *The Strange Theory of Light and Matter*.

4. See T.S. Eliot's essay on Andrew Marvell, *Selected Essays*, 25.

5. No, I haven't read Jacques Lacan. But there's a handy little book called *Postmodernism for Beginners*, by Richard Appignanesi and Chris Garratt, which helps keep track of the various strands of thought that make up the complex web of postmodernism. In fairness, the idea that language creates thought does have deep historical roots. In his book on chemistry, *The Same and Not the Same*, Roald Hoffmann quotes from the eighteenth-century Abbé de Condillac: "We think only through the medium of words," 66.

6. Stephen Pinker discusses the Sapir-Whorf hypothesis and comments, "The idea that Eskimos pay more attention to varieties of snow because they have more words for it is so topsy-turvy (can you think of any other reason why Eskimos might pay attention to snow?) that it's hard to believe it would be taken seriously were it not for the feeling of transcending common sense," 124ff.

7. See Wigner, "The Unreasonable Effectiveness of Mathematics in the Natural Sciences."

8. See "Letter to a Japanese Friend" in Wood and Bernasconi, 1–5.

9. Maryanne Wolfe's book, *Proust and the Squid*, is an illuminating discussion of what it takes for the brain to master reading.

10. See Pinker, 14.

11. See Miall, 109.

12. Ibid., 108.

13. This is the major theme of Damasio's *Descartes' Error*. The case of Elliott, the former stockbroker, is discussed on page 34ff.

14. See Miall, 100.

15. Ibid., 137.

16. This eighteenth-century haiku by Issa is translated by Yukari Meldrum. Kisikata, as it is known today, is a coastal town in the Japanese

prefecture of Akita. Traditionally in literature, the place name is Hisagata.

17. These are the first two lines of the title sequence in Mimi Khalvati's collection, *The Meanest Flower*.

18. See Miall, 112.

FIVE Symmetry

1. See Møller and Thornhill, 174–92.

2. You can get a sense of the approaches, theories and complexities of symmetry perception in *Human Symmetry Perception and its Computational Analysis*, edited by C.W. Tyler.

3. Donald Coxeter's quote is cited in Roberts, 100.

4. The difference between how we handle symmetrical patterns in seeing vs. hearing is discussed in the paper by J. Wagemans, "Detection of visual symmetries," in Tyler, 26.

5. Robert Pinsky's short book, *The Sounds of Poetry*, is an excellent analysis of how meter and stress patterns interact in English poetry, while Robert Jourdain discusses the parallel relationship in music: "Meter is tyrannical in its regularity, but it is sure. In contrast, phrasing largely works through the inherent meaning of the sound it contains. Just as a spoken phrase is finished when something meaningful has been said, musical phrases carry an entire musical idea," 131.

6. See Jourdain, 45.

7. This does not apply to the polyrhythmic patterns of African music, where more than one beat is happening simultaneously.

8. Jourdain discusses how critical rhythm is to melody on page 81 of *Music, the Brain, and Ecstasy*.

9. Ibid., 81.

10. Jourdain's discussion of how children learn tunes can be found on page 61ff.

11. See Jourdain, 80.

12. Ibid., 147.

13. See Levitin, 25.

14. Thanks to composer Don Ross for explaining to me how a deceptive cadence works.

15. John Milton uses this phrase in his preface to *Paradise Lost*.

16. See Pinsky, 81.

17.   Randall, 240.

18.   A number of popular physics books explain the emergence of four fundamental forces after the Big Bang. Gribbin, 335ff.

19.   See Jourdain, 126.

20.   See Tsur, "Rhyme and Cognitive Poetics."

21.   Ibid., 5.

22.   Charles Olson's essay on projective verse can be found in Geddes, 603.

23.   Adam Bradley discusses issues of style and originality in hip hop, in Chapter 4 of his *Book of Rhymes*.

24.   Ellen Dissanayake discusses this motivation extensively in *Art and Intimacy*.

25.   Lines from Coventry Patmore's long poem *The Angel in the House* (from the section titled "The Paragon").

26.   Reuven Tsur discusses the "markedness" of dactylic meter extensively in "Rhyme and Cognitive Poetics."

27.   Antonio Damasio outlines the somatic marker hypothesis in *Descartes' Error*. Robert Jourdain's speculation that somatic markers may be caused by rhythm can be found on page 327 of *Music, the Brain, and Ecstasy*.

28.   Jourdain, 329.

29.   In *The Emperor's New Mind*, Roger Penrose describes the development of aperiodic tilings (173–78) their five-fold symmetry and their relation to quasicrystals (562ff).

30.   The first lines of "Ulysses", by Alfred, Lord Tennyson.

31.   See Tsur, *Toward a Theory of Cognitive Poetics*, 156.

32.   See Roberts, 99.

33.   See Magueijo, 156.

SIX   Poetry and Scale

1.   Benoit Mandelbrot's famous quote can be found in the introductory paragraph of *The Fractal Geometry of Nature*, and goes on to say, "compared with Euclid—a term used in this work to denote all of standard geometry—Nature exhibits not simply a higher degree but an altogether different level of complexity."

2.   The carpet and its fractal dimension are illustrated and described in Mandelbrot, 144. As the process of cutting out smaller and smaller squares continues, the carpet's area actually vanishes and the total perimeter of its holes becomes infinite.

3. Ibid., 7.

4. Ibid., C2 and C16.

5. Natural Resources Canada.

6. This is the fractal dimension of the west coast of Great Britain, originally calculated by Lewis Fry Richardson and cited by Mandelbrot, 26ff.

7. See Dawkins, 485.

8. Brian Goodwin briefly outlines the centrality of self-referential networks to both nature and culture in his contribution to Brockman, ed., 171.

9. See Mandelbrot, 113.

10. See Taylor et al., "Fractals: A Resonance Between Art and Nature?"

11. See Jourdain, 132.

12. See Taylor et al., "Fractal Expressionism."

13. See Ladefogend, 87.

14. Tsur discusses the acoustic coding of vowels in *Toward a Theory of Cognitive Poetics*, 216ff.

15. See Taleb, "The Roots of Unfairness: The Black Swan in Arts and Literature."

16. See Schroeder, 187ff.

17. Lisa Randall discusses this phenomenon (known as "T-duality") in her book *Warped Passages*. "Odd as it may seem, in string theory, extremely small and extremely large rolled-up dimensions yield the same physical consequences," 450–51.

SEVEN   The Ultraviolet Catastrophe

1. A number of books about the history of quantum physics cover the "ultraviolet catastrophe" and its implications for Max Planck's decision to treat light as a collection of quantized particles. See Farmelo, 6–12; Gribbin, 221–22; Aczel, 34–35; and Penrose, 502–03.

2. See Lightman, 43–59, for an accessible summary of the development and significance of Einstein's historic paper and for the text of the paper itself.

3. Chapter 10 ("Violet") of Victoria Finlay's fascinating book *Colour* describes the discovery of the aniline dyes.

4. See Genz, 81.

5. The clash with Eddington is outlined in Arthur Miller's biography of "Chandra," *Empire of the Stars*.

6.   There are many descriptions of the characteristics of black holes, including Brian Greene's excellent *The Fabric of the Cosmos*, 477ff.

7.   See Murdoch, 96. Her discussion of tragedy continues through the following chapter.

8.   See Chapter 6 ("Grief") of Melvin Konner's excellent book, *The Tangled Wing*.

9.   Oliver Sacks describes achromatopsia and communities where it occurs at an unusually high rate in *The Island of the Colorblind*.

10.   Catastrophe theory, its development by René Thom and its implications, are explained with exceptional clarity in *Mathematics and the Unexpected*, by Ivar Ekeland.

11.   The human perception of blue is discussed in Peter Pesic's book on how scientists have struggled for centuries to answer that simple question, "Why is the sky blue?" See Pesic, 161–70.

12.   See Krause, 44.

EIGHT   Gather Ye Rosebuds

1.   Don Marquis first made this observation in his column in the *New York Sun*, *The Sun Dial*.

2.   Steven Pinker discusses the "semantic fussiness" of verb constructions extensively in Chapter 2 of *The Stuff of Thought*.

3.   Dr. Johnson's remark is cited in Shlain, 93.

4.   See Gell-Mann, 17. Gell-Mann's book explores the science of complexity.

5.   Lee, "Notes on Rhythm in Poetry."

6.   The poetic features of motherese are discussed extensively in Miall and Dissanayake, "The Poetics of Babytalk."

7.   Ibid.

8.   Usha Goswami summarizes research on rhyme and early childhood development in *The RoutledgeFalmer Reader in Psychology of Education*, edited by Harry Daniels and Anne Edwards.

9.   Untermeyer, ed., 264.

10.   See Pinsky, 115.

11.   See Eliot, "Tradition and the Individual Talent," in *Selected Essays*, 10.

12.   See Olson, "Projective Verse," in Geddes, ed., 603.

13.   The poem "Healing Words" is found in Shirley Serviss's collection, *Hitchhiking in the Hospital*.

14.   See Brodsky, 211.

15. See Ritchie, "Roses in Winter"; special thanks to Dr. Neville Arnold for pointing me to this article.

16. In the microfiche filing cabinet up on the second floor of the Edmonton Public Library, in envelope 3600, you can find two microfilm negatives covered with tiny images of the pages of *Voix de la solitude*. It was a book of poetry published in 1938, with five hundred copies printed by Les Éditions du Totem of Montreal. Clearly, poetry print runs haven't changed much since then.

17. From "certain maxims of archy" by Don Marquis in *Archy and Mehitabel*, 53.

NINE Motion

1. Much of this section is drawn from Nicholas Harberd's fascinating book, *Seed to Seed: The Secret Life of Plants*. See Harberd, 178 (plants as collections of tubes); 69 (the meristem); 90–91 (navigation system for roots); 93 (meristem genes); 37–38 (a map being made and read at the same time); and 155 (floral meristem).

2. See Grant, "The Trembling Giant."

3. This is the thesis of *Who Murdered Chaucer?*, by Terry Jones et al. The authors piece together a painstaking amount of research to explore Chaucer's final years.

4. See Dawkins, 549.

5. An excellent general source of information about the cerebellum (and other structures of the brain) is Dubuc, *The Brain from Top to Bottom*, on McGill University's website.

6. See Levitin, 170–71.

7. The Australian Broadcasting Corporation covered the 2009 flooding of Lake Eyre, in its May 17 news item, retrieved from http://www.abc.net.au/news/video/2009/05/17/2572897.htm

8. For the arsenal of brain systems that pigeons use, see Blechman, Chapter 10.

9. See Mithen, 247.

10. The estimated age of "mitochondrial Eve" is discussed in Cavalli-Sforza and Cavalli-Sforza, 68.

11. See Dawkins, 416.

12. See Cavalli-Sforza and Cavalli-Sforza, 56.

13. The dating of humans in Brazil is discussed in "Pedra Furada, Brazil: Paleoindians, Paintings and Paradoxes."

14. In their paper "The Revolution that Wasn't," Sally McBrearty and Alison S. Brooks challenge the idea of a radical change in human behaviour at the so-called "bottleneck," pointing out that it may be a Eurocentric bias and failure to appreciate the archeological record in Africa.

15. See Hurford, "Human Uniqueness, Learned Symbols and Recursive Thought."

16. Hauser, Chomsky and Fitch suggest that recursion may be *the* defining feature of the "narrow" faculty of language, i.e., the development that sets human speech apart from other animal thought processes. However, Hurford and others argue that elements of recursive thought are found in other species, especially those that have to keep track of dominance hierarchies.

17. Loreto Todd's brief but informative book, *Pidgins and Creoles*, points out that pidgins and creoles are syntactically more similar to each other than to the languages from which their lexicons derive. Todd, 28.

18. See Johannes Krause et al., "The Derived FOXP2 Variant of Modern Humans Was Shared with Neanderthals."

19. See Chatwin, 13.

20. See Aporta, "The Trail as Home."

21. Hellum, 30.

22. You can find the Proto-Indo-European lexicon on the website of the Linguistics Research Center at the University of Texas at Austin: http://www.utexas.edu/cola/centers/lrc/ielex/PokornyMaster-X.html

TEN   Politics as Phase Space

1. The quotation comes from James Gleick's classic book, *Chaos*. Pages 134–40 have an excellent introduction to the concept of phase space.

2. See Stewart, *Life's Other Secret*, 111–12.

3. See the excellent biography, *The Life and Times of Chaucer*, by novelist John Gardner.

4. See Jones et al., *Who Murdered Chaucer?*

ELEVEN   Brain Surgery

1. The anecdote about Hamilton and Wordsworth is documented in Stewart, *Why Beauty is Truth*, 139–40.

2. Yes, I spelled it "hare."

3. See Frost, 487.

4. See Snyder, "Explaining and Inducing Savant Skills."

5. The quote from Robert Graves can be found in Geddes, 561.

6. See Randall, 339.

7. It's difficult to determine at this stage just what does malfunction in schizophrenia. The usual figure cited for the frequency of schizophrenia is 1 per cent of the population, although an analysis by Dinesh Bhugra indicates the figure may be more like four people per thousand.

8. See Sternberg, 144, for the discussion of habituation, and also Akdag et al., "The Startle Reflex in Schizophrenia."

9. The role of the bard is described by Osborn Bergin in "Bardic Poetry."

10. See West, 30.

11. Gerald Edelman's quote from *Bright Air, Brilliant Fire* is also cited by David Lodge in *Consciousness and the Novel*, in an excellent discussion of lyric poetry as "arguably man's most successful effort to describe qualia," 10ff.

Bibliography

Aczel, Amir D. *Entanglement: The Greatest Mystery in Physics*. Vancouver: Raincoast Books, 2002.

Alighieri, Dante. *The Portable Dante*. Edited and translated by Mark Musa. New York: Penguin Classics, 2003.

Akdag, S.J. et al. "The Startle Reflex in Schizophrenia: Habituation and Personality Correlates." *Schizophrenia Research* 64, no. 2–3 (November 15, 2003): 165–73.

Aporta, Claudio. "The Trail as Home: Inuit and their Pan-Arctic Network of Routes." *Human Ecology* 37, no. 2 (2009): 131–46.

Appignanesi, Richard, and Chris Garratt. *Postmodernism for Beginners*. Cambridge, UK: Icon Books, 1995.

Aristotle. *Nichomachean Ethics*. Translated by W.D. Ross. Retrieved from *The Internet Classics Archive* (MIT). Accessed June 13, 2011, http://classics. mit.edu/Aristotle/nicomachaen.2.ii.html

Barnes, Allison, and Paul Thagard. "Empathy and Analogy." Waterloo, ON: University of Waterloo, 1997. Accessed June 13, 2011, http://cogsci. uwaterloo.ca/Articles/Pages/Empathy.html

Bekenstein, Jacob D. "Information in the Holographic Universe." *Scientific American* (August 2003), 61.

Bergin, Osborn. "Bardic Poetry: A Lecture Delivered in 1912." Retrieved from the *Corpus of Electronic Texts*, University College Cork. Accessed June 13, 2011, http://www.ucc.ie/celt/bardic.html

Bhugra, Dinesh. *The Global Prevalence of Schizophrenia*. In *Public Library of Science*, published online May 31, 2005. Accessed July 31, 2011, http://www.plosmedicine.org/article/info%3Adoi%2F10.1371%2Fjournal.pmed.0020151

Blakeslee, Sandra, and Matthew Blakeslee. *The Body has a Mind of Its Own*. New York: Random House, 2008.

Blechman, Andrew. *Pigeons*. New York: Grove Press, 2006.

Boland, Eavan. *Object Lessons: The Life of the Woman and the Poet in our Time*. New York: W.W. Norton & Company, 1995.

Borges, Jorge Luis. *Selected Non-fictions*. New York: Viking, 1999.

Bradley, Adam. *Book of Rhymes: The Poetics of Hip Hop*. New York: BasicCivitas/Perseus Books, 2009.

Brockman, John, ed. *What We Believe but Cannot Prove: Today's Leading Thinkers on Science in the Age of Certainty*. New York: Harper Perennial, 2006.

Brodsky, Joseph. "An Immodest Proposal." In *On Grief and Reason*. New York: Farrar, Straus and Giroux, 1995.

Browning, Elizabeth Barrett. *Aurora Leigh*. London: The Women's Press, 1985; originally published in 1857.

Budiansky, Stephen. *The Character of Cats: The Origins, Intelligence, Behavior, and Stratagems of Felis silvestris catus*. New York: Penguin, 2003.

Bugnet, Georges. *Voix de la solitude*. Montreal: Les Éditions du Totem, 1938.

Carruthers, Peter, and Andrew Chamberlain, eds. *Evolution and the Human Mind: Modularity, Language and Meta-cognition*. Cambridge, UK: Cambridge University Press, 2000.

Capra, Fritjof. *The Tao of Physics*. London: Fontana Paperbacks, 1976.

Cavalli-Sforza, Luigi Luca, and Francesco Cavalli-Sforza. *The Great Human Diasporas: The History of Diversity and Evolution*. New York: Basic Books, 1995.

Chandrasekhar, S. *Truth and Beauty: Aesthetics and Motivations in Science*. Chicago: University of Chicago Press, 1990.

Chatwin, Bruce. *The Songlines*. New York: Penguin, 1981.

Damasio, Antonio. *Descartes' Error*. New York: Avon Books, 1994.

———. *The Feeling of What Happens: Body and Emotion in the Making of Consciousness*. San Diego, CA: Harcourt, 1999.

Daniels, Harry, and Anne Edwards, eds. *The RoutledgeFalmer Reader in Psychology of Education*. London: RoutledgeFalmer, 2004.

Dawkins, Richard. *The Ancestor's Tale: A Pilgrimage to the Dawn of Life*. London: Phoenix, 2005.

Decety, Jean, and Claus Lamm. "Human Empathy through the Lens of Social Neuroscience." *The Scientific World Journal* 6 (2006): 1146–63.

Dissanayake, Ellen. *Art and Intimacy: How the Arts Began*. Seattle: University of Washington Press, 2000.

Dobyns, Stephen. *Best Words, Best Order: Essays on Poetry*. Second edition. New York: Palgrave Macmillan, 2003.

Dubuc, Bruno. *The Brain from Top to Bottom*. McGill University website. Accessed June 13, 2011, http://thebrain.mcgill.ca/flash/d/d_06/d_06_cr/d_06_cr_mou/d_06_cr_mou.html

Ekeland, Ivar. *Mathematics and the Unexpected*. Chicago: University of Chicago Press, 1988.

Eliot, T.S. *Selected Essays*. New York: Harcourt, Brace & World, 1960.

Farmelo, Graham, ed. *It Must be Beautiful: Great Equations of Modern Science*. London: Granta Publications, 2002.

Feynman, Richard P. QED: *The Strange Theory of Light and Matter*. Princeton, NJ: Princeton University Press, 1985.

Finlay, Victoria. *Colour: Travels through the Paintbox*. London: Hodder and Stoughton, 2002.

Frost, Robert. *Collected Poems, Prose, and Plays*. Edited by Richard Poirier and Mark Richardson. New York: Library of America, 1995.

Gardner, John. *The Life and Times of Chaucer*. New York: Alfred A. Knopf, 1977.

Gardner, Martin. *Relativity for the Million*. New York: Macmillan Company, 1962.

Geddes, Gary, ed. *20th Century Poetry and Poetics, 3rd Edition*. Toronto: Oxford University Press, 1985.

Gell-Mann, Murray. *The Quark and the Jaguar: Adventures in the Simple and the Complex*. New York: W.H. Freeman and Company, 1994.

Genz, Henning. *Nothingness: The Science of Empty Space*. Cambridge, MA: Perseus Publishing, 1999.

Gibbs, Raymond W. Jr., ed. *The Cambridge Handbook of Metaphor and Thought*. New York: Cambridge University Press, 2008.

Gleick, James. *Chaos: Making a New Science*. New York: Penguin, 1987.

Grant, Michael C. "The Trembling Giant." *Discover Magazine* (October 1993). http://discovermagazine.com/1993/oct/thetremblinggian285

Greene, Brian. *The Elegant Universe: Superstrings, Hidden Dimensions and the Quest for the Ultimate Theory*. New York: Random House, 1999.

———. *The Fabric of the Cosmos: Space, Time and the Texture of Reality*. New York: Random House, 2004.

Gribbin, John. *In Search of the Big Bang: Quantum Physics and Cosmology*. New York: Bantam New Age Books, 1986.

Gwynn, R.S., ed. *New Expansive Poetry: Theory, Criticism, History*. Ashland, OR: Storyline Press, 1999.

Harberd, Nicholas. *Seed to Seed: The Secret Life of Plants*. London: Bloomsbury Publishing, 2006.

Hauser, Marc D., Noam Chomsky, W. Tecumseh Fitch. "The Faculty of Language: What is It, Who has It, and How Did It Evolve?" *Science* 298, no. 5598 (November 22, 2002): 1569–79.

Hellum, A.K. *Listening to Trees*. Edmonton, AB: NeWest Press, 2008.

Hoffmann, Roald. *The Same and Not the Same*. New York: Columbia University Press, 1995.

Hofstadter, Douglas. "Analogy as the Core of Cognition." In *The Best American Science Writing 2000*. Edited by James Gleick. New York: Ecco Press, 2000.

Holub, Miroslav. *The Rampage*. Translated by David Young et al. London: Faber & Faber, 1997.

Hurford, James R. "Human Uniqueness, Learned Symbols and Recursive Thought." *European Review* 12, no. 4 (October 2004): 551–65.

Johnson, Steven. *Everything Bad is Good for You: How Today's Popular Culture Is Actually Making Us Smarter*. New York: Riverhead Books, 2006.

Jones, Terry, Robert Yeager, Terry Dolan, Alan Fletcher and Juliette Dor. *Who Murdered Chaucer?: A Medieval Mystery*. New York: St. Martin's Press, 2003.

Jourdain, Robert. *Music, the Brain, and Ecstasy: How Music Captures Our Imagination*. New York: HarperCollins, 2002.

Khalvati, Mimi. *The Meanest Flower*. Manchester: Carcanet Press, 2007.

Koch, Kenneth. *Wishes, Lies and Dreams: Teaching Children to Write Poetry*. New York: Harper Perennial, 1999. Reprint.

Konner, Melvin. *The Tangled Wing: Biological Constraints on the Human Spirit*. New York: Henry Holt and Company, 2002.

Krause, Johannes, et al. "The Derived FOXP2 Variant of Modern Humans Was Shared with Neanderthals." *Current Biology* 17 (November 2007) 1–5. Accessed August 2, 2011, http://www.sciencedirect.com/science/article/pii/S0960982207020659

Krauss, Lawrence M. *The Physics of Star Trek*. New York: Harper Perennial, 1996.

Ladefogend, Peter. *Vowels and Consonants: An Introduction to the Sounds of Language*. Oxford: Blackwell Publishing, 2001.

Lee, Dennis. "Notes on Rhythm in Poetry." Reprinted in *Thinking and Singing: Poetry and the Practice of Philosophy*. Edited by Tim Lilburn. Toronto: Cormorant Books, 2002.

Levitin, Daniel J. *This is Your Brain on Music: The Science of a Human Obsession*. New York: Penguin, 2006.

Lightman, Alan. *The Discoveries: Great Breakthroughs in 20th Century Science*. New York: Vintage Books, 2005.

Lodge, David. *Consciousness and the Novel: Connected Essays*. Cambridge, MA: Harvard University Press, 2002.

Lutz, Tom. *Crying: The Natural and Cultural History of Tears*. New York: W.W. Norton & Company, 1999.

Magueijo, João. *Faster than the Speed of Light: The Story of Scientific Speculation*. Cambridge, MA: Perseus Books, 2003.

MacEwen, Gwendolyn. *Magic Animals: Selected Poetry of Gwendolyn MacEwen*. Toronto: Stoddart, 1984.

Mandelbrot, Benoit B. *The Fractal Geometry of Nature*. Revised edition. New York: W.H. Freeman and Company, 1983.

Marquis, Don. *Archy and Mehitabel*. New York: Anchor Books, 1973.

McAdam, Rhona. *Cartography*. Lantzville, BC: Oolichan Books, 2006.

McBrearty, Sally, and Alison S. Brooks. "The Revolution that Wasn't: A New Interpretation of the Origin of Modern Human Behavior." *Journal of Human Evolution* 39, no. 5 (November 2000): 453–563.

Miall, David S. *Literary Reading: Empirical and Theoretical Studies*. New York: Peter Lang, 2006.

Miall, David S., and Ellen Dissanayake. "The Poetics of Babytalk." *Human Nature* 14, no. 4 (2003): 337–64.

Miller, Arthur I. *Empire of the Stars: Obsession, Friendship and Betrayal in the Quest for Black Holes*. Boston: Houghton Mifflin, 2005.

Milosz, Czeslaw, ed. *A Book of Luminous Things: An International Anthology of Poetry*. San Diego, CA: Harcourt Brace and Company, 1996.

———. *The Witness of Poetry*. Cambridge, MA: Harvard University Press, 1984.

Mithen, Steven. *The Singing Neanderthals*. London: Orion Books, 2005.

Møller, A.P., and R. Thornhill. "Bilateral Symmetry and Sexual Selection: A Meta-Analysis." *The American Naturalist* 151, no. 2 (February 1998): 174–92.

Murdoch, Iris. *Metaphysics as a Guide to Morals*. London: Penguin, 1993.

Natural Resources Canada. *Atlas of Canada*. Accessed June 13, 2011, http://atlas.nrcan.gc.ca/site/english/learningresources/facts/coastline.html

Neruda, Pablo. *Selected Poems: A Bilingual Edition*. New York: Penguin, 1975.

Page, P.K. *Hologram: A Book of* Glosas. London, ON: Brick Books, 1994.

Pagels, Heinz. *The Cosmic Code: Quantum Physics as the Language of Nature*. New York: Bantam New Age Books, 1984.

Papagno, Costanza, Arianna Fogliata, Eleonora Catricala and Carlo Miniussi, "The Lexical Processing of Abstract and Concrete Nouns." *Brain Research* 1263 (March 2009): 78–86.

Patmore, Coventry. *The Angel in the House*. 1891. The Victorian Web. http://www.victorianweb.org/authors/patmore/angel/

"Pedra Furada, Brazil: Paleoindians, Paintings and Paradoxes (An interview with Drs. Niède Guidon, Anne-Marie Pessis, Fabio Parenti, Claude Guérin, Evelyne Peyre, and Guaciara M. dos Santos)." *Athena Review* 3, no. 2 (March 2002). Accessed July 31, 2011, http://www.athenapub.com/10pfurad.htm

Penrose, Roger. *The Emperor's New Mind*. London: Oxford University Press, 1989.

———. *The Road to Reality*. London: Vintage Books, 2005.

Pesic, Peter. *Sky in a Bottle*. Cambridge, MA: MIT Press, 2007.

Pinker, Steven. *The Stuff of Thought: Language as a Window into Human Nature*. New York: Viking, 2007.

Pinsky, Robert. *The Sounds of Poetry: A Brief Guide*. New York: Farrar, Straus and Giroux, 1998.

Ramachandran, V.S., and Sandra Blakeslee. *Phantoms in the Brain: Probing the Mysteries of the Human Mind*. New York: William Morrow and Company, 1998.

Randall, Lisa. *Warped Passages: Unravelling the Mysteries of the Universe's Hidden Dimensions*. New York: Harper Perennial, 2005.

Ritchie, Gary A. "Roses in Winter." *American Rose* (Jan–Feb 2009), 23.

Roberts, Siobhan. *King of Infinite Space: Donald Coxeter, the Man Who Saved Geometry*. Toronto: House of Anansi Press, 2006.

Sacks, Oliver. *The Island of the Colorblind*. Toronto: Random House, Vintage
    Canada Edition, 1998.

Schroeder, Manfred. *Fractals, Chaos, Power Laws: Minutes from an Infinite
    Paradise*. New York: W.H. Freeman and Company, 1991.

Serviss, Shirley A. *Hitchhiking in the Hospital*. Edmonton, AB: Inkling Press,
    2005.

Shlain, Leonard. *Art and Physics*. New York: Harper Perennial, 2007.

Smolin, Lee. *Three Roads to Quantum Gravity*. New York: Basic Books, 2002.

Snyder, Allan. "Explaining and Inducing Savant Skills: Privileged Access to
    Lower Level, Less-processed Information." *Philosophical Transactions of
    the Royal Society* B, no. 364 (2009): 1399–05.

Sternberg, J. *A Handbook of Creativity*. Oxford: Oxford University Press, 1999.

Stewart, Ian. *Why Beauty is Truth: A History of Symmetry*. New York: Basic
    Books, 2007.

———. *Life's Other Secret: The New Mathematics of the Living World*. New York:
    John Wiley & Sons, 1998.

Taleb, Nicholas Nassim. "The Roots of Unfairness: The Black Swan in Arts
    and Literature." *Literary Research/Recherche Littéraire*, Journal of the
    International Comparative Literature Association, 2004. Accessed
    August 2, 2011, http://www.fooledbyrandomness.com/ARTE.pdf

Taylor, Richard, et al. "Fractals: A Resonance Between Art and Nature?"
    Accessed June 13, 2011, http://vismath7.tripod.com/proceedings/taylor.
    htm

Taylor, Richard, Adam P. Micolich and David Jonas. "Fractal Expressionism:
    Can Science be Used to Further Our Understanding of Art?" *Physics World*
    (October 1999).

Todd, Loreto. *Pidgins and Creoles*. New York: Routledge, 1990.

Tsur, Reuven. *Toward a Theory of Cognitive Poetics, Second Edition*. Brighton:
    Sussex Academic Press, 2008.

———. "Rhyme and Cognitive Poetics." *Poetics Today* 17 (1996): 55–87. Accessed
    July 10, 2011, http://cogprints.org/735/1/RhymeGestalt_2.html

Turner, Mark. *The Literary Mind: The Origin of Thought and Language*. Oxford:
    Oxford University Press, 1996.

Tyler, C.W., ed. *Human Symmetry Perception and its Computational Analysis*.
    Mahwah, NJ: Lawrence Erlbaum Associates, 2002.

Untermeyer, Louis, ed. *The Letters of Robert Frost to Louis Untermeyer*. New
    York: Holt, Rinehart and Winston, 1963.

Walcott, Derek. *Omeros*. London: Faber & Faber, 1990.

West, M.L. *Indo-European Poetry and Myth*. Oxford: Oxford University Press, 2007.

Whitman, Walt. *Notes and Fragments*. London: Ardley Press, 2008.

Wigner, Eugene. "The Unreasonable Effectiveness of Mathematics in the Natural Sciences." *Communications in Pure and Applied Mathematics*, vol. 13, no. I (February 1960). New York: John Wiley & Sons, Inc.

Williamson, Margaret. *Sappho's Immortal Daughters*. Cambridge, MA: Harvard University Press, 1995.

Wolfe, Maryanne. *Proust and the Squid: The Story and Science of the Reading Brain*. New York: Harper Perennial, 2008.

Wood, David, and Robert Bernasconi, eds. *Derrida and Différance*. Warwick: Parousia Press, 1985.

Index

schizophrenia, 226–27, 228,
    249n7
See also Alzheimer's disease
metaphors, 17–46
    analogies and thinking, 19–20,
        24–25
    brain research on, 28–29, 41–42
    changes in perception from,
        44–45
    clichés, 36
    comparison of sensory fields in,
        32–34
    conceptual metaphors, 25–26
    consciousness and, 27–28
    cross-cultural metaphors, 26, 34
    dead metaphors, 31
    emotions and, 29–30
    hyperboles, 31–32
    language and, 17–18, 24–27,
        29–30
    Mark Turner on, 18–19
    mathematical equations and,
        32–38, 41
    metonymy and, 18
    parataxis and, 42–43
    parts of speech and, 25, 30–31
    phase spaces as, 210–12, 248n1
    play as, 17–18
    politics and, 44
    proverbs and, 19
    puns and, 31–32
    as riddles, 34–35
    sensory perceptions and, 38–40
    similes and, 40–42
    "small spatial stories" and,
        18–20
    as social lubricant, 31

synodoches, 18
therapy and, 44
zeugma, 240n3
Miall, David, 81, 84, 91, 176, 180,
    246n6
migration, 196–98
Milton, John, 105, 123
monkeys, vervet, 201–02
Montgomery, Lucy Maud, 50,
    58–60
mothering. *See* child development
motion
    brain structures and, 194–95
    human purpose and, 206–07
    plant and animal processes and,
        192–95
Mozart, Wolfgang Amadeus,
    219–20, 221
Mungo, Lake, Australia, 199
Murdoch, Iris, 153–54
muscular dystrophy, 158–60
muscular system
    brain system and, 194–95
    human evolution, language and,
        202
    muscle, etymology of term,
        26–27
    somatic marker hypothesis,
        118–19
music
    child prodigies, 219–20
    comparison with poetry, 91,
        98–101, 243n5
    completeness in, 102
    human evolution and, 202
    melody in, 98–102
    rhythm in, 98–99, 110, 133